THE PROTOTYPING METHODOLOGY

THE PROTOTYPING METHODOLOGY

Kenneth E. Lantz

Prentice-Hall
A Division of Simon & Schuster, Inc.
Englewood Cliffs, New Jersey

ISBN 0-8359-5897-3

Published by Prentice-Hall
A Division of Simon & Schuster, Inc.
Englewood Cliffs, New Jersey 07632

10 9 8 7 6 5 4 3 2 1

Printed in the United States of America.

CONTENTS

Chapter Eight

PREFACE

Two of the most severe problems facing Information Systems groups today are the increasing backlog of service requests and the decline in user confidence. These problems, in turn, are eroding the influence of Information Systems within their organizations. The proper use of prototyping will alleviate these problems and consequently enhance the influence of Information Systems.

Numerous articles about prototyping have appeared recently. Most people think they know what it means. Many organizations have tried it, but prototyping is not easy. If you have attempted it, you know this is true. Most organizations are not receiving the full benefits of it. Many Information Systems managers are rightly leery of using it without a methodology.

The purpose of this book is to explain what prototyping is and to provide a methodology for using it. This book should help you whether you are a programmer, analyst, project leader, manager, director, vice president, consultant, or student of Information Systems methodologies. It should also help you if you are a user of Information Systems services who wants to learn about prototyping.

If you are deciding whether or not to prototype, this book is for you.

If you have prototyped, or are prototyping, and want to improve your approach, this book is for you.

If you want to learn how to prototype, this book is for you.

If you only want to learn *about* prototyping, this book is for you, too.

THE PROTOTYPING METHODOLOGY

Chapter One

WHAT PROTOTYPING IS AND WHY IT IS VALUABLE

WHAT PROTOTYPING IS

Although the word "prototyping" has not made it into the dictionary yet, Webster has several definitions for the noun "prototype." The last[1] applies best.

> *prototype: a first full-scale and usually functional form of a new type or design of a construction (as an airplane)*

We in Information Systems are not the first to use the word "prototype," anymore than we are the first to prototype. As Webster notes, our friends in aerospace build prototypes, and they started long before we did.

The subject of this book is "software prototyping," not all forms of prototyping. After applying words carefully in the definition, this book will normally use the simple term prototyping, as is commonly done in the Information Systems literature.

The definition:

> *Software prototyping is an information system development methodology based on building and using a model of a system for designing, implementing, testing, and installing the system.*

Prototyping Is a Methodology

Table 1-1 shows the important features of a methodology. Since a methodology is a collection of methods, you may organize the methods into steps. You may write them down in the order in which they should be executed. You may teach both the steps and the order in which they should be done.

TABLE 1-1 Features of a Methodology

A Methodology Is Something You Should Be Able:

- to teach
- to schedule
- to measure
- to compare
- to modify

When you are using a methodology, you may estimate the time and resources that you will need for each step in an application of the methodology and for any project, as a whole, on which you will use it. You may prepare a schedule from these estimates.

When you are managing projects on which a methodology is used, you may measure progress against the schedule and the plan.

You may compare the use of a methodology on one project with its use on other projects, and you may compare its use by one group to its use by other groups. You may also compare one methodology against other methodologies intended for similar purposes.

You may also modify one or more parts of a methodology based on your experience with it and your measurements and comparisons of its uses.

Prototyping is a methodology; it is a collection of methods. It is done in a systematic way, and that way can be described (the premise of this book).

Many in Information Systems management have viewed prototyping as a "quick and dirty" approach talented people use for urgent systems projects. These people often seem to be "winging it," when they're prototyping, taking shortcuts that eliminate necessary steps in the system development process. For this reason, Information Systems managements are often cautious about using it. Even if they used it once or twice on critical projects, it was not systematic enough for them to use it generally. They want a methodology, for all the advantages that a methodology brings.

Prototyping Is Building a Model

"What do you really mean by prototyping?" people often ask. "Do you *actually mean* doing such things as setting up the database and the screens for updating it?" The answer must be an emphatic "Yes." Prototyping is based on building a model of the system to be developed. Moreover, the initial model should include the major program mod-

TABLE 1-2 Contents of Initial Prototype Model

- Major Program Modules
- Database
- Screens
- Reports
- Inputs and Outputs for Interfacing Systems

ules, the database, screens, reports, and the inputs and outputs that the system will use for communicating with other (interfacing) systems.

If you want to design and build a model of something, you must first learn about the thing you want to model. An assumption of prototyping is that you don't need to know the full requirements of a system in order to build the model. In fact, an important product of prototyping is the completed requirements. But you must know enough about the requirements to build a useable initial model.

Using the Model for Designing the System

Prototyping uses the model for designing the system. How can the model be a means for designing the system? When the designer designed the prototype, he designed the system, didn't he? No, he only designed the prototype. Now that prototype will become the vehicle for designing the full system.

The initial version of the prototype is not the full system, but it does contain its designer's understanding of the database, screens, and reports. As the user and Information Systems people begin to work with the prototype, it will change.

Also, remember the initial version of the prototype is a skeletal version of the system; it does not contain all the processing and validation rules that the system will ultimately have. As those working with the prototype modify it and add to it, they will be completing the design of the system. So, the prototype will be a vehicle for designing the final version of the system.

Using the Model for Implementing the System

Prototyping uses the model to implement the system. In today's average organization, you write programs in some problem solution language, such as COBOL, to implement a system.

The initial version of the prototype will consist of programs written in some language (possibly a fourth generation language) to move data back and forth between screens, the database, reports, and the inputs and outputs used to communicate with interfacing systems. At first, these programs may do little processing; they may actually dummy it. For example, a program that produces data for a report may generate the data from "hard coding" instead of extracting it from the data base.

As the prototyping process continues, newer versions of programs, that perform more closely to those of the ultimate system, will replace the original versions. For example, a program that actually extracts data for a report from a database may replace one that dummied out data. Originally, you may write most versions of the prototype's programs in fourth generation languages; later you may rewrite at least some of them in the language you have chosen for the production version of the system.

Using the Model for Testing the System

Prototyping uses the model to perform both the system and the acceptance testing of the system. The initial version of the prototype, as well as all subsequent versions, will communicate with system test versions of feeding systems and systems to be fed. So, the prototype will always run in "system test" mode. And, since the user will be working with the prototype from the beginning, the user will be performing "acceptance testing" of the prototype from the beginning.

Using the Model for Installing the System

Prototyping uses the model for installing the system. Those who must learn how to work with the production system as part of the installation process—normally, users— will learn how to use it by working with the prototype. The prototype produces material, such as sample screens and reports, which is useful for the documentation done as part of the system installation. Also, part of the prototyping of a system is the prototyping of any conversion process required for installing the system.

The prototype becomes the system We have not consistently used the concept of a prototype in Information Systems. Often, when we use the word prototyping, we are really talking about building and using mock-ups. A mock-up, according to Webster, is different from a prototype. Webster defines a mock-up[2] as follows:

mock-up: a full-sized structural model built accurately to scale chiefly for study, testing, or display

The mock-up of a system is a skeletal version of it; from the outside the mock-up generally behaves like the system itself eventually will. Inside it may be doing little more than moving data from program stub to program stub. The aerospace and automobile industries use mock-ups for such things as wind-tunnel testing. In Information Systems, we use mock-ups for such things as testing the structure of a new system.

The definition of a mock-up, as it applies to developing systems, follows.

A software mock-up is a system development approach based on building and using a model of a system for discovering the requirements of the system.

Prototyping does not go beyond a mock-up if, after building and experimenting with the initial model of the system, and possibly making a few modifications to it, you abandon it when you have gained a good understanding of the requirements. Following this approach, you primarily use the model, often built with fourth generation languages, to help the user visualize the system better, and to verify and to expand on the basic requirements. But instead of expanding the model into a full prototype, you use more traditional means to document what the full system should do, then build the production version of the system.

A prototype as a mock-up is valuable. Nonetheless, it does not provide all the benefits of a full prototype, as discussed below.

With the Prototyping Methodology, even though a prototype may be little more than a mock-up when it is first built, it becomes the first one of its kind by the time it is finished. So, when the prototyping process ends, the prototype has become the system.[3]

THE PROTOTYPE BECOMES THE SYSTEM

PROTOTYPING METHODOLOGY

The first question to be answered for the Prototyping Methodology or any other systems development methodology, such as the currently popular "life cycle" methodologies, is what does the methodology require that you do? Table 1-3 lists those things which must be a part of any systems development methodology for it to be complete.

TABLE 1-3 Components of Systems Development Methodology

- Study Present System
- Determine Feasibility of New System
- Define Requirements
- Design
- Program
- Test
- Acceptance Test
- Train Users
- Convert from Old System
- Install

If the present system is to be studied, Information Systems professionals usually do it. The present system may be manual or a combination of manual and office machine processes. In most of today's medium to large size organizations, systems also contain some computerized processes.

The aim of any study of the present system should be to discover the data that is processed, filed, and used in the system, the volumes of data that are processed, the timing of the system's processing, how it communicates with other systems, and any problems with its operation.

Before the decision is made to begin a project for a new system, Information Systems people, often working with people from other components of the organization, may determine the project's feasibility. This determination includes understanding the business problem to be addressed by the proposed system, doing a preliminary design of the new system, estimating the time and other resources that would be required to implement it, examining alternatives, and deciding whether and when to proceed.

Defining the requirements of a system is a process of thinking through what a system must do and communicating the results of that thinking to those who will design the system.

Designing a system is determining how to meet the system's requirements. It may be done in steps: the first step is to define the major system's processes and to identify the data to be processed; the second step is to do the physical design of the system's programs and database.

Programming a system includes determining in detail (specifying) how the individual parts of the system will work, translating the system into machine-readable programs, and performing the unit testing of these programs.

Testing is the system testing of the communication between the different parts of a system and between the system and the other systems with which it communicates. The basis of system testing should be a plan that describes what is to be tested and how the testing will be done. System testing is first done with relatively small amounts of test data. After the system has run successfully with it, the system is tested with a full volume of data.

Those who will use a new system (the users) acceptance test it to ascertain whether it meets their requirements. After a successful acceptance test, the user usually "signs off" the system so that it may be "put into production."

Those who will use the new system must be taught how to use it. If there is to be any user documentation, it should be written ahead of time, so they may use it in their training.

A new system rarely starts from scratch. Even if the old system is entirely manual, it probably contains data that must somehow be converted for use in the new system. If the data is in, or can be put into, machine-readable form, programs may be written to convert the data to the new system's database.

Installing a system includes developing production copies of all the programs in the system, writing and testing instructions to the operating system, and documenting the system for the computer operations staff and those who will be maintaining it. Installing a system should also include establishing a permanent test version of the system.

Classic Life Cycle versus Prototyping

The plan for a classic, non-prototyping, "life cycle" style project should look something like that shown in Figure 1-1.

Determine Feasi-bility	Study Present System	Define Require-ments	Design New System	Program	Test	Accep-tance Test	Train Users	
							Convert	Install

FIGURE 1-1 Typical "life cycle" project plan

In a "life cycle" methodology, the various project phases usually follow each other serially. About the only overlapping that is done is training the users at the same time as converting to the new system.

The "life cycle" methodologies emphasize scheduling and meeting milestones, producing documentation at the milestones and getting signoffs on it, no signoff being more important than the user's signoff of the requirements. These methodologies may lead us to view the Information Systems development process as a set of rigorously distinct phases performed at separate times. Each phase depends on the immediately preceding phase, and those doing the work may not begin the next one before the previous one has been completed. It is difficult for those who are used to the "life cycle" methodologies to imagine beginning programming before the system design has been completed. It is even more difficult for them to think of beginning system testing before the programming has been finished. Moreover, even if they could imagine these things, they would never do them.

Those of us who have designed and implemented information systems recall often feeling a strong impulse to iterate what we were doing. Maybe we felt we should somehow repeat at least part of the phase we were working on. Perhaps we thought we should re-examine things that had been determined in earlier phases. Following the "life cycle" methodologies has helped[4] us. But, the "life cycle" methodology, especially the milestone dates that we had promised to meet, kept us from succumbing to the temptation to iterate. After all, if we ever started, HOW WOULD WE KNOW WHEN TO STOP ITERATING?

The plan for a project following the Prototyping Methodology should resemble Figure 1-2..

One thing that is different about the prototyping plan above is the overlapping of most of the project phases with each other. This implies that prototyping provides the iteration which is often sought in developing Information Systems. Actually, prototyping provides more than an iteration of phases; it provides a continuation of phases.

The requirements definition begins after the present system has been studied. Before the requirements definition is completed, the design begins. Soon after it has begun, the programming begins. Shortly after that the testing, acceptance testing, and user training begin. The milestone dates for completing the requirements definition, design, programming, testing, and acceptance testing are usually the same date. The completion of user training is probably the same date as the completion of conversion. Installation begins after that.

This does not mean that a feature of the system is designed before there are requirements for it. Nor does it mean that something is programmed before it has been designed. And how could something be tested before it was programmed? What it does mean is that things

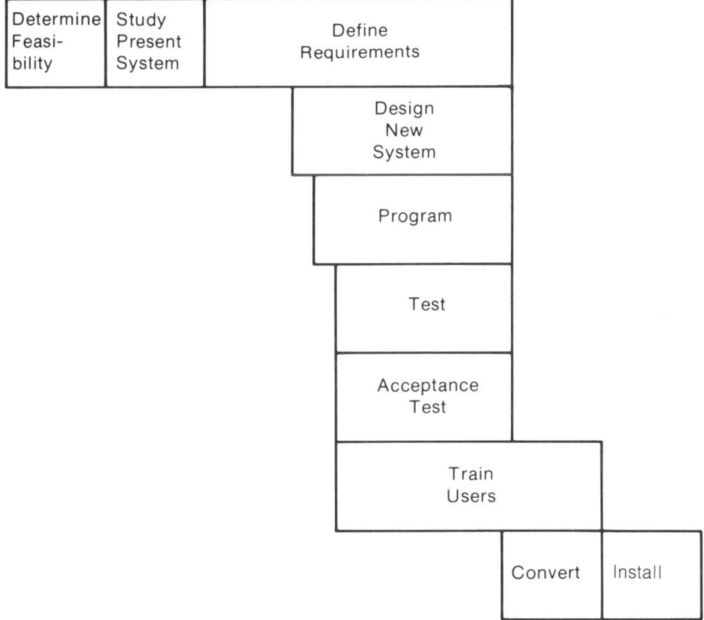

FIGURE 1-2 Typical prototyping project plan

learned during acceptance testing influence the requirements for additional parts of the system, and subsequently the design, programming, testing, acceptance testing, and training. Changes in the design certainly affect the programming. The programming affects the testing. Everything that goes before affects the user, who is learning the system as he does the acceptance testing. So, in a sense, the work of these phases continues until they all have been completed.

Much of the work of defining requirements will be finished before most of acceptance testing is done. Likewise, much of the design will have been completed, before most of the system testing is done. Yet, things discovered during system testing or user training or acceptance testing can and should affect requirements definition and design and programming.

Later chapters of this book will provide more detail on the different phases of the Prototyping Methodology, but a few additional comments here may preclude some misunderstandings.

Even though the two figures were not drawn to scale, if you compare the "Prototyping Methodology" to "How It Is Usually Done" you may conclude that using the Prototyping Methodology will require less calendar time for a project than following a "life cycle" methodology. That's an accurate conclusion.

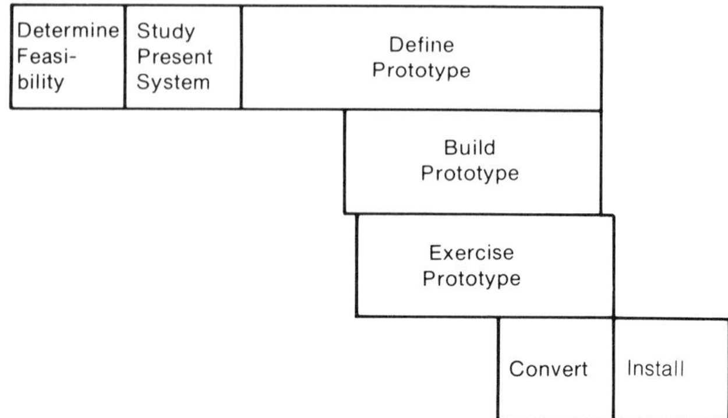

FIGURE 1-3 New way of looking at prototyping project approach

If you compare the figures further you may notice that the individual requirements, design, program, test, acceptance test, and user training phases will take longer under the Prototyping Methodology than under a "life cycle" methodology. That's also an accurate conclusion. But you should also consider that because they overlap each other, they will take less calendar time together than they will in the "life cycle" methodology.

Because some of the phases are longer with the Prototyping Methodology, you might conclude that the Prototyping Methodology will require more manhours from Information Systems people. Actually, the Prototyping Methodology should require less manhours from Information Systems people. But, it will require more manhours from the user than he is likely to expend with a "life cycle" methodology.

Figure 1-3 might imply that the Prototyping Methodology is simplistic and does not include all the things that must be a part of any systems development methodology. That is not true. Prototyping is streamlined, but it is not simplistic. The figure emphasizes the streamlined aspect of the Prototyping Methodology.

The streamlined phases are:

- Define Prototype
- Build Prototype
- Exercise Prototype.

Table 1-4 relates the prototyping phases to the traditional phases.

TABLE 1-4 *Prototyping versus Traditional Project Phases*

PROTOTYPING PHASE	INCLUDES	TRADITIONAL PHASE
Determine Feasibility		Determine Feasibility
Study Present System		Study Present System
Define Prototype		Define Requirements
Build Prototype		Design
Exercise Prototype		Program Test Acceptance Test Train Users
Convert		Convert
Install		Install

WHY PROTOTYPE?

Table 1-5 lists the principal advantages of prototyping.

TABLE 1-5 Advantages of Prototyping

- Pleases users
- Reduces development cost
- Decreases communication problems
- Lowers operations costs
- Slashes calendar time required
- Produces the right system the first time
- Cuts manpower needed

Prototyping pleases users. Instead of only going to requirements meetings and walkthroughs, reviewing screen and report layouts, and signing off various documents, many of which he may not fully understand, the user actively participates in the design and development of the system on a day-to-day basis. Prototyping makes him feel far more involved with the new system than the "life cycle" methodology could ever make him feel. He can exercise the system everyday. If the system doesn't meet his needs, he can request modifications and see them quickly.

Prototyping reduces the total cost of system development. According to Gremillion and Pyburn,[5] "A number of companies report that total systems development costs by the prototype approach are usually less than 25% of the costs with the traditional approach."

Prototyping decreases communication problems, not only between user and Information Systems people but also between those working on the new system being prototyped and those working on systems it must communicate with.

Since the user actively participates in the design and development of the system, he does not have the severe communication problems that sometimes occur when other methodologies are used. Also, since the prototype itself, not paper, is the main medium of communication of system functions between the user and Information Systems people, they misunderstand each other much less. People can tell you what they don't like about an existing system more easily than they can tell you what they want in a system you are designing on paper.

Systems development plans usually leave the system testing of a new system until the very end of the project. And the plan never seems to allow enough time for correcting problems, if system testing does not go well. Needless to say, problems often do occur, and fixing them adds an embarrassing delay to installing a new system. In contrast, the Prototyping Methodology requires that, from the first, the prototype be "brought up" in system test mode, so it may communicate with any systems that feed it or any that it must feed. System interface problems should be solved as soon as they are found, because the prototype may not run properly until they are corrected. Also, as prototyping progresses, if any modifications to the prototype impair its communications with interfacing systems, they will be detected and quickly corrected.

Prototyping lowers operations costs. Systems developed by prototyping use less computer resources. This advantage is the opposite of what many people expect; they think all systems developed by prototyping contain at least some parts written in a fourth generation language, making the systems computer resource "hogs." But the use of fourth generation languages is not a necessary part of prototyping, as is discussed later. Certainly, if operational efficiency is a criterion for a system and if the use of fourth generation languages would prevent it, a system developed through prototyping would not contain programs written in them. Remember, though, that fourth generation languages are becoming more efficient day by day.

So, leaving the inappropriate use of fourth generation languages aside, how are systems developed through prototyping more efficient in the use of computer resources than systems developed in other ways? They produce less reports and screens programmed to meet the "I think I need a..." requirement. Users often think they need a report or screen, which they never use after the system is installed. Systems developed through prototyping do not usually produce such reports and screens. Also, systems developed through prototyping contain fewer validations and unnecessary controls. Users realize what it will cost and the time it will take to add validations that might be nice, but which they do not feel are necessary, when they learn how the system will operate. Also, Information Systems people do not add controls that they might have otherwise added, because they did not know how the user would work with the system.

Prototyping slashes the calendar time required to complete a project. Since the work of several phases is done concurrently, a project done with the Prototyping Methodology should always take less time than one done with the "life cycle" methodology, unless significantly fewer resources are used with the Prototyping Methodology. For example, if a project that would require ten people with a "life cycle" methodology is only staffed with two with the Prototyping Methodology, it will probably require more calendar time with prototyping, despite the overlapping of phases that prototyping provides.

Prototyping produces the right system the first time. It would be difficult for it to do otherwise, with the active participation of the user in the system design and development and the constant running of the prototype with interfacing systems. Also, the requirements that are the basis of the sytem are not theoretical; they are developed from actual experiences with the prototype.

Prototyping cuts the manpower needed for a project compared to what would be required using a "life cycle" methodology. Table 1-6 lists the factors that influence this.

Excessive documentation is not needed with the Prototyping Methodology as it may be with a "life cycle" methodology. Because communications are often not good between users and Information Systems people, documenting the system in great detail seems to be good business practice. Some view the reason for such documentation as protection for Information Systems. Although essential documentation is a part of the Prototyping Methodology, the work required to produce such excessive documentation does not need to be done.

TABLE 1-6 How Prototyping Methodology Cuts Manpower Needs

- Doesn't require excessive documentation
- Good communication takes less time than poor communication
- No last-minute surprises requiring major modifications
- Much of user training a by-product not a separate phase

Good communication requires less time than poor communication. Often "life cycle" review and status meetings seem to get off the track. Documentation and correspondence often seem to confuse the recipients. When people who are supposed to be working together don't understand one another, it seems to take extra time to communicate simple information. But when people are working together effectively on a project they all understand, much more is communicated in much less time.

The constant, incremental development and testing (both system and acceptance), which are an important part of prototyping, keep any one problem from becoming a major consumer of time. Also, no surprises requiring major reworking occur during testing. How many times, using traditional methodologies, has a flaw been discovered that required major modifications before a system could be installed? The likelihood of finding such flaws is much less with the Prototyping Methodology.

Much of the training of key users occurs as a by-product of the prototyping process, not as part of a separate phase.

Advantages of Full Prototype Versus Mock-up

The differences in the advantages of a full prototype compared to a prototype as a mock-up are important to consider. Table 1-7 shows some of the major differences. You should remember that, since most of what are called "prototypes" in current Information Systems literature are actually "mock-ups," many of the reasons given for prototyping in that literature apply only to using mock-ups.

Possible Disadvantages

Table 1-8 lists some possible disadvantages of using the Prototyping Methodology.

TABLE 1-7 Advantages of Full Prototype vs Mock-up

ADVANTAGE	FULL PROTOTYPE	PROTOTYPE AS MOCK-UP
System meets user's needs	X	X
User participates in design	X	X
Early discovery of design problems	X	X
Requirements are developed experientially not theoretically	X	X
User participates in development	X	
System interface problems must be solved from the start	X	
Work of several phases concurrent with development	X	
Excessive documentation not needed	X	X
Unexpected problems that consume major amounts of time don't occur	X	
Much user training a by-product of development	X	

TABLE 1-8 Possible Disadvantages of Prototyping Methodology

- Visible use of computer resources
- Object system may be less efficient
- Requires cooperation between user and Information Systems
- Some view prototyping as an art not a methodology

Using prototyping development tools may expend more computer time than traditional development approaches. This may concern Information Systems managements, since the use of computer resources is the easiest to measure of all their uses of an organization's resources. The tradeoff facing them is the use of computer time versus the use of Information Systems professionals.

If the production version of the system performs much processing with code from fourth generation languages, it may require more computer time for running than if it were written in a language like COBOL. (It's ironic that only a few years ago many were lamenting that programs written in COBOL were less efficient that ones written in assembly language.) Certainly, if processing efficiency is an important design criterion, all of the production system probably should not be written in a fourth generation language. Nothing in the Prototyping Methodology requires the use of fourth generation languages for the production versions of systems. Also, other tradeoffs may affect the relative efficiency of the production system, such as development cost versus cost of computer resources and loss of opportunity cost versus cost of computer resources.

Stating that the cooperation between user and Information Systems that prototyping requires is a possible disadvantage may seem sarcastic. Yet in some organizations the relationships between Information Systems and users have deteriorated to the point that any project requiring cooperation would be doomed from its beginning. Prototyping offers a chance, but no guarantee, to turn such situations around.

If clever Information Systems people prototype as a way of expressing their artfulness in implementing systems, the organization will benefit by however much work those talented people can do. But their management may rightly worry about using prototyping when it is an art not a methodology.

What's Wrong with Current Methodologies

Wetherbe describes the current state of Information Systems well[6] when he declares that the major problems with our current approaches are that they take too long, and they don't produce the right system the first time.

Why are we following approaches that serve us so poorly? Until recently we have been mainly building batch systems. The machine time we needed to build and test them was expensive, compared to the costs for the people doing the building and testing. Furthermore, we were not able to compile and test frequently.

Information Systems people need a lot of information to design application systems. They know hardware and systems design, but often they are not experts in the functions of their organization. They want to do a complete job, covering all possible situations that may ever occur in the systems; their desire to be thorough may even lead them to over-design. Nevertheless, when they have finished, users often view the systems that they have produced as far from what was wanted.

On the other hand, no matter how well users understand the functions of their areas of the organization, and no matter how creative they may be, they often have difficulty defining what they want a new system to do. So, Information Systems frequently perceives them as not knowing what they want.

In far too many cases the situation between Information Systems and the users has become a modern day "range war" between the good guys and the bad guys.

In the opinion of many observers of the Information Systems scene, the use of a "life cycle" methodology, establishing information centers, and promoting the use of personal computers have been the best solution, so far, to this war.

Many adherents of the "life cycle" methodology consider the writing and formal signoff of detailed requirements as its principal benefit. Also, these requirements protect Information Systems, if the final system conforms to them. *But regardless of what the requirements describe and no matter who signed them off, no user is happy when the system does not do what he had hoped it would.*

You all know the story of far too many projects done using current methodologies. If they are not cancelled, they are often completed late and over budget. Usually they are "done in" by one or more surprises discovered during system testing or acceptance testing. Even though iteration was not planned, the projects must be halted for modifications based on new information. And how many times does a project leader discover that the "new" project he has been assigned to is actually being attempted again for the third or the fourth time, after having been cancelled all the previous times?

So, the backlog, both visible and invisible, of requests for Information Systems projects continues to grow despite, or possibly in some sense, because of, current methodologies.

In many organizations, Information Systems is no longer considered responsive to users' needs. Users may be beginning to go around Information Systems to satisfy their information needs. Methodologies that do not promote user confidence have a slim chance, at best, to help change the deteriorating situation.

Table 1-9 lists some of the principal problems with current methodologies.

TABLE 1-9 Problems with Current Methodologies

• Are thorough	BUT DON'T	• please users
• Produce extensive documentation	BUT DON'T	• decrease communication problems
• Identify project steps	BUT DON'T	• decrease calendar time needed
• Describe system thoroughly	BUT DON'T	• guarantee it's the right system
• Delineate skills needed	BUT DON'T	• cut manpower needed
• Track project costs	BUT DON'T	• reduce them

APPLICATION OF PROTOTYPING

Where should prototyping be used? On what types of projects? Some would limit the use of prototyping to the development of systems that are poorly defined. Others[7] would limit it to ones that are relatively small. In fact, some[8] claim that the use of prototyping leads to a tendency to keep projects smaller and thus more manageable. Others[9] believe that prototyping applies best to simulating human interaction with systems; whereas using other methodologies, such as structured programming, for the parts of the system with which people do not communicate directly, will increase productivity. Certainly, many will tell you, prototyping should not be used for installing software packages.

In some cases where prototyping might otherwise be recommended, its use might not be a good choice. If prototyping violates important criteria for the project, it shouldn't be used. If, for example, one criterion for implementing a minor modification to a system is to complete it in two weeks, and just getting a prototype of the unmodified system running would take more than a week, then, all other things being equal, prototyping would probably not be a good choice for the project. Chapter 2 describes when to prototype in more detail. Nevertheless, prototyping has all those uses shown in Table 1-10.

Prototyping is useful in developing on-line systems. Most people writing about using prototyping would agree with this. Moreover, on-line systems are even difficult for systems designers to visualize if they are only working on paper. Using some kind of prototyping, at least for the flow of information to and from screens, is a natural solution to this problem for many designers. Many of them started prototyping when they realized the screen mapper that was a part of a fourth generation language they had acquired would help them. All of the advantages of prototyping apply to its use for on-line systems.

Prototyping uses a model to design and develop a batch system the same as it does for an on-line system. But its usefulness for batch

TABLE 1-10 Applications of Prototyping Methodology

- On-line systems
- Batch systems
- Packages
- Enhancements and modifications

systems is difficult for many people to understand. As was said above, it's a "natural" for on-line systems; they're a part of the new technology, as is prototyping. They go together. Batch systems are different. After all, we've been doing batch systems for years; by now we certainly know how to do them. But, how do we do them? If we're advanced, we use some type of "life cycle" methodology.

Using the "life cycle" methodologies has the disadvantages that were discussed earlier; using prototyping has the advantages discussed earlier. But you still may not see how prototyping applies to batch systems. After all, batch systems don't have any screens. Nevertheless, the same problems arise in developing batch and on-line systems:

- Users have difficulty visualizing what they want a system to do.
- Users have difficulty evaluating requirements documentation.

Users have trouble visualizing what should be on a report, but, as you may have experienced many times, they know what they don't like when they see a report. Information Systems people often misunderstand the best sequence for data. Processing rules are often based on incorrect information about the data to be used. Users often realize, after working with a system, that reports they felt were essential, are actually unnecessary. Outputs a system produces for groups other than the primary user are often not useful, at least, not without changes. Prototyping solves these problems and provides all its other advantages.

Prototyping is useful for systems based on software packages. It also provides extra value, if it is used in the package selection process. Nevertheless, it is still valuable, even when it is not introduced until the installation of the package.

A package is either installed with or without modifications. Package modifications include using options or exits provided by the package and adding pre-processors and post-processors to the package. Building a prototype that consists of the package with its modifications should provide all the advantages of prototyping. You may doubt that this approach will decrease calendar time or cut project costs, but have you considered how many times adjustments have had to be made to a system based on a package, immediately after its supposedly smooth installation? Even if no modifications are planned to a package, installing it first as a prototype with test data will enable you to test it, to ensure it meets the user's needs, to train key users, to ensure that the user and Information Services people understand it, and to determine whether it meshes well with interfacing systems.

Prototyping also enhances the package selection process. Usually, a package selection reduces to a choice between two or three strong candidates. But, even if there is only one strong candidate that appears to be a shoo-in, prototyping will help you make a good choice. Most package vendors allow a trial use before purchase or, at least, before a purchase is final. Use it. Installing a prototype of the package and a model of any modifications you plan to make will help both the user and Information Systems people to evaluate the package's strengths and weaknesses and how it meets the requirements.

Prototyping is useful for enhancements and modifications to existing systems. If a test system is already available for the system to be changed, prototyping will be that much easier. Even if a test system does not exist, you only need to prototype that part of the system that you are planning to change. Of course, you must determine whether prototyping would be practical, particularly for relatively small projects, as described in the first part of this section.

If an enhancement or modification is to be relatively minor, bringing up a prototype of it should be relatively minor. After all, you should intend to test the change before it becomes part of the production system. Prototyping a small change is essentially only system testing and acceptance testing it; you should do these things in any case.

If a change is larger, the benefits of prototyping and its applicability are the same as for new systems, including eliminating communication problems, pleasing the user, and helping you to install the right change the first time.

WHAT IS NEEDED TO PROTOTYPE

Sarvari lists[10] such things as on-line technology, a relational database, and a high level language as required for prototyping. Others mention[11] the need for such things as a database management system, data dictionary, query language and report writer.

Do you really need fourth generation languages? Report writers? DBMSs? Data Dictionaries? Prototyping is certainly easier with them than it is without them. And, these aids have generally been available to people when they began prototyping.

Just having fourth generation languages does not guarantee that you can prototype. And using them does not necessarily mean that you are prototyping. Many organizations that have all those aids do not prototype or, if they do, they do not feel that their prototyping is productive. Also, giving too much attention to these aids may cloud an important issue: what makes prototyping work?

TABLE 1-11 What Is Needed To Prototype

- Understanding the data
- Moving data through the system and to and from connecting systems
- Exercising the prototype frequently
- Building and modifying the prototype quickly

The essential response is that proper project management and the use of a methodology make prototyping work. No matter how good the tools available to Information Systems people are, good project management is essential to their successful use of the tools. Also, a methodology makes prototyping a tool and not just an art.

But, in addition to these essential elements, what is it at the nuts and bolts level that makes prototyping work? Table 1-11 lists the four principal ingredients.

You need to understand the data to design and build the initial prototype model. That initial prototype model will consist of:

- Major program modules
- Database
- Screens
- Reports
- Inputs and outputs for interfacing systems

You don't want to produce screens and reports that don't show data realistically. Also, prototyping will go more smoothly if you base the initial version of the database on a good understanding of the data—for example, if it is within 75 % of the content and structure of the production version. Finally, the structure of the initial model itself depends heavily on the database, screens, reports, inputs, and outputs; and, they depend on an understanding of the data. Chapter 4 will discuss ways to gain an understanding of the data.

Moving data through the system and to and from connecting systems provides many of the principal benefits of prototyping. It makes both systems and acceptance testing work by showing what happens when you enter input data and when you modify its processing. The feeling of realism it gives also promotes user training.

Since the core of prototyping is using the model, the user and Information Systems people are performing its most essential function as they work together exercising it. How often to exercise the prototype is a matter of judgement. If too much time elapses between prototyping sessions, your experience with the prototype gets stale, you may need additional time to refresh participants' memories, and you may begin documenting more elaborately. These things may rob you of many of the advantages of prototyping.

Being able to build and to modify the prototype quickly is important for exercising it frequently. But the speed required for modifications is relative to other factors. If, for example, different things are being done in five successive exercise sessions, no one may want to see modifications resulting from the first session until after the end of the fifth session. Chapter 6 discusses exercising the prototype in more detail.

Tools To Enhance Prototyping

Tools that will help you prototype are tools that will help you understand and move the data, and exercise and modify the prototype. None of them may be essential, but using at least some of them will help you to prototype.

No specific set of tools is best for prototyping. Using one data dictionary may not be better than using another. Writing programs with one fourth generation language may not be better than writing them with another. But using a data dictionary may be better than manually controlling information about the data. And using a fourth generation language may be better than coding everything in COBOL.

TABLE 1-12 Tools To Enhance Prototyping

NEEDED TO PROTOTYPE	TOOLS TO HELP
• Understanding the data	• Data dictionary
• Moving the data	• Interactive testing
	• Test data generator
	• Library control of program modules
• Exercising the prototype	• Interactive testing
	• Test data generator
• Modifying the prototype	• Interactive testing
	• Library control of program modules
	• Fourth generation language (Report Writer, Screen Painter, Query Language)
	• Relational DBMS

A data dictionary should help you to record and to organize information about data. This should aid you in understanding the data.

An interactive testing system allows you to use a terminal to change programs quickly, to submit frequent tests, and to review the results of test sessions without having to wait for batch outputs. Using interactive testing helps you to exercise the prototype often, to modify it quickly, and thus to move data between its components.

A test data generator produces test data for your use in moving data through the system and exercising the prototype.

Although it may seem obvious, library control of program modules aids in making quick changes to the prototype and in moving data between system components by providing such things as version control of programs and database definitions, and multiple uses of the same program (may be a feature of the operating system).

Quickly modifying those parts of a prototype that you have written in a fourth generation language should be easy to do. The features of the fourth generation language that are particularly helpful are the report writer, screen painter and query language. You may obtain some or all of these separately without getting a fourth generation language. One consideration I must add is that as you add more features to the prototype, programming them in a fourth generation language may become more complex than in a problem solution language. I will review this question in more detail in Chapter 5.

A relational data base management system, either alone or as part of a fourth generation language, may help you in modifying your database design, especially in the early stages of prototyping.

SUMMARY

1. Definition of prototyping: Software prototyping is an information system development methodology based on building and using a model of a system as the means of designing, implementing, testing, and installing it.
2. Prototyping
    - Is a methodology
    - Is building a model
    - Uses the model for designing the system
    - Uses the model for implementing the system
    - Uses the model for testing the system
    - Uses the model for installing the system
3. The prototype becomes the system.

4. A prototype is the first fully developed version of a system; a mock-up is a skeletal version of it.
5. Prototyping is iterative; the "life cycle" methodologies are serial.
6. Advantages of prototyping:
   - Pleases users
   - Reduces development cost
   - Decreases communication problems
   - Slashes calendar time
   - Produces the right system the first time
   - Cuts manpower needs
7. Possible disadvantages of prototyping:
   - Its use of computer resources is easy to measure
   - Object system it produces may be less efficient
   - It requires cooperation between user and Information Systems
   - Some claim it is an art not a methodology
8. Current methodologies:
   - Take too long
   - Don't produce the right system the first time
9. Prototyping is useful for:
   - On-line systems
   - Batch systems
   - Packages
   - Enhancements and modifications
10. Elements needed to make prototyping work:
    - Good project management
    - Use of methodology
    - Understanding the data
    - Moving data through the system and to and from connecting systems
    - Exercising the prototype frequently
    - Building and modifying the prototype quickly
11. Tools to enhance prototyping:
    - Data dictionary

- Interactive testing
- Library control of program modules
- Fourth generation language (Report Writer, Screen Painter, Query Language)
- Relational DBMS

FOOTNOTES

1. *Websters New Collegiate Dictionary*, Springfield MA: G. & C. Merriam Co., 1977
2. Ibid.
3. Chapin, Ned,"Prototyping- Quick, Not Dirty ..." *Data Management*, 1983, 21 (10), 46-48
4. Gilhooley, Ian, "System Development Mythology" *Datamation*, 1983, 29 (6), 272-276
5. Gremillion, Lee L. & Pyburn, Philip,"Breaking the Systems Development Bottleneck" *Harvard Business Review*, 1983, 61 (2), 133
6. Wetherbe, James C., "Systems Development: Heuristic or Prototyping;" *Computerworld*, 1982, 16 (17), SR14-15
7. Meyer, K.& Kovaco, A., "Model Systems," *Datamation*, 1983, *29* (9) 248,250,252
8. Ibid.
9. Frank, Werner L., "Structured vs. Prototyping Methodology" *Computerworld*, 1983, 17 (33), 51-52
10. Sarvari, I.L., "The Case for Prototyping" *Canadian Datasystems*, 1983, 15 (10), 100-104
11. Marett, W., "New Development Tool Project Based On 'C' " *New Zealand Interface*, July 1983, 16-17

Chapter Two

ESTABLISHING A DEVELOPMENT PROJECT

This chapter describes the project initiation process for new systems and system changes to be done using the prototyping methodology. The project initiation process includes:

- Initiating and processing service requests
- Evaluating alternatives
- Deciding whether to establish a project
- Planning the next phases.

This chapter discusses the essential parts of a service request, the steps that should be taken to process it, how to consider alternatives, the role prototyping should play, factors the evaluation should include, information that should be captured, understanding the effects of charge-backs and Information Systems planning committees, and planning the next phases.

FEASIBILITY DETERMINATION AND SERVICE REQUESTS

Service requests are formal requests for the assistance of Information Systems. Their requestors prepare service requests because they want something developed, either a new system, a change, or an addition to an existing system. Some organizations also use service requests to report problems, but that use of them will not be discussed here. The service request explains to Information Systems what the requestors need and why they need it.

The requestors of service requests are usually users—persons from outside the Information Systems organization who want the assistance of Information Systems. But Information Systems may also use them to authorize a project for a system listed in the information

TABLE 2-1 Uses for Service Requests

- Give visibility to Information Systems' work
- Ensure user management reviews them
- Ensure user needs are understood
- Determine availability of information systems professionals
- Evaluate against the data strategy
- Make sure of fit with functional systems plan
- Estimate time and costs
- Determine their feasibility
- Assign them a priority
- Secure approval for them
- Control (not lose) them

systems plan or a database needed to satisfy the organization's data strategy.

Information Systems organizations use Information Systems service requests for several reasons. Table 2-1 lists them.

Give visibility to information systems' work All components of an organization serve the goals of the whole organization. Manufacturing or flight operations produce the product of the organization. Purchasing buys the materials the organization needs. Marketing and sales sell the product. Personnel serves the employees' and employee administrative needs. Accounting records and reports on the financial state of the organization. Members of the organization generally do not question the value of these components, yet they often do question the contribution made by communications and Information Systems.

Communications and Information Systems only serve the other components of the organization. Other persons in the organization may not recognize what they do as valuable. So, Information Systems should record and inform people what it is doing for the organization.

Ensure user management reviews them Often individual service requests only represent a need of the manager who submits them. They may not meet the needs of his upper management nor support his organization's long-range plans. In fact, they may conflict with them. For example, the manager of Payroll may want five new reports to assist his people in performing their work, but the Director of Accounting, to whom he reports, may have agreed with the Vice President for Administration on a study for a new, integrated payroll-personnel system. If the Director of Accounting does not have the opportunity to review the request from Payroll, work may be initiated at first only to be scrapped later.

Ensure user needs are understood Information Systems should understand users' needs and the business problems they are trying to solve in order to produce systems or changes that promote the goals of the organization and are worth the money and manpower expended on them. Processing service requests in a systematic and orderly way helps Information Systems to learn more about the users' needs.

The use of service requests may slow Information Systems' response. This seems frustrating, particularly where users are enthusiastic users of Information Systems. Users today would like to pick up the phone to ask for a system, and receive word next morning that it is up and running. Information Systems people sometimes feel apologetic that they cannot provide such service. But, too often, responding too quickly produces poorly designed, truly unresponsive systems and system modifications.

Developing a problem statement, as described below, helps ensure that the right problems are solved by Information Systems' response to the service request.

Determine availability of information systems professionals In any organization that has had computer systems for more than six months, the time of Information Systems professionals is scarce. Whether the number and types of professionals who will be needed would be available must be considered. The request may need to have a high priority to ensure the necessary resources will be assigned to complete it on time.

Evaluate against the data strategy Information Systems should not begin work on a service request before it has been evaluated against the organization's data strategy. A data strategy identifies what data is essential for operating the organization, who is responsible for creating and maintaining it, how it is to be maintained and communicated, and who may use it.

Make sure of fit with functional systems plan Information Systems should also evaluate requests in the light of the organization's functional systems plan. A functional systems plan includes the operations to be performed by each system, the system's functional characteristics (such things as volumes of data to be processed and required response times), its relationships to other systems, and its relative importance.

Estimate time and costs Information Systems and the initiating users should estimate the calendar time and costs that may be incurred to satisfy the service request.

Determine their feasibility Information Systems should recommend whether the work requested in a service request is feasible. This feasibility recommendation is based on whether the benefits of the work outweigh the estimated costs of developing and operating the requested system. This used to be the principal focus of the processing of service requests. Now, the availability of Information Systems professionals and the organization's data strategy and functional systems plans also play important roles.

Assign them a priority Information Systems will recommend a relative priority for any service request whose approval it recommends. This recommendation is based on the estimated benefits of the service request compared to its costs, its need for scarce resources, and the date it has been requested. If the organization has an Information Systems planning committee composed of senior representatives from various components of the organization, this committee may assign the final priority.

Secure approval for them Information Systems management should approve them. Also, the management of the initiating user group should approve them. If the organization has an Information Systems planning committee, it should also approve them. Each approver must consider the request in the light of the functional system and data strategies of the organization, the time required by systems professionals, whether it appears feasible, and its effect on the other plans of the organization.

Control (not lose) them The controls that Information Systems establishes for service requests will normally ensure that they are not lost. This may seem a trivial use of controls, but in too many organizations that do not use them, one main frustration of users is not knowing what has become of their requests and whether they will ever receive a response to them.

SERVICE REQUEST FLOW

Information Systems logs an incoming service request to establish a control on it and assigns a business systems analyst to review it. The analyst evaluates the request and records his recommendation on the service request.

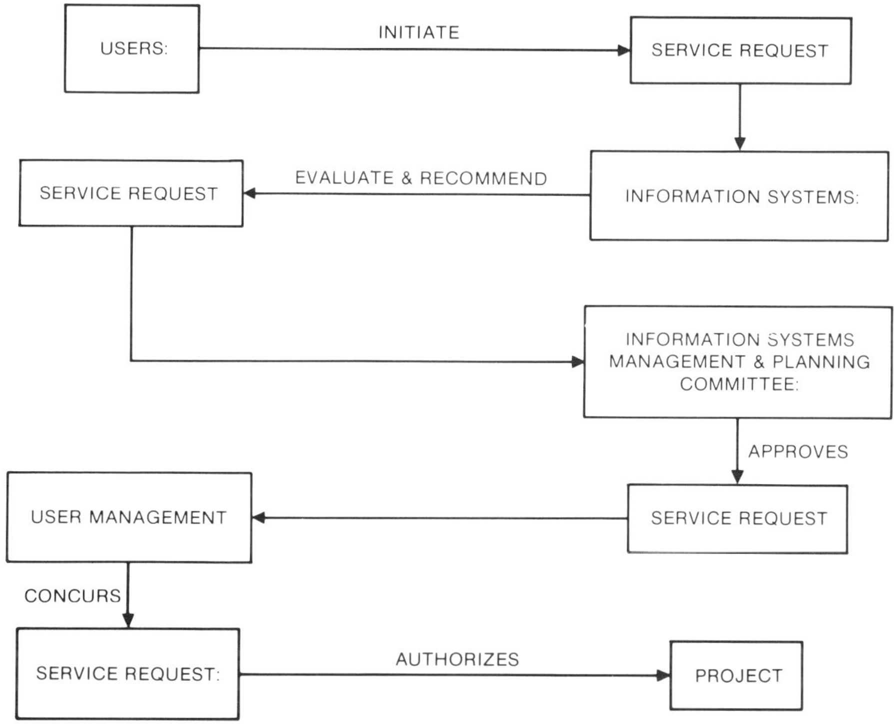

FIGURE 2-1 *Service request information flow*

The management of Information Systems, and any Information Systems operations planning committee established to evaluate requests for Information Systems work, countersign the recommendation and establish the priority of any work to be done.

If the management of the requestor of the service request concurs in the recommendation, the service request becomes the authorization for establishing a project.

PARTS OF A SERVICE REQUEST

Table 2-2 lists the basic parts of a service request. Figure 2-2 shows a sample form. The parts of a service request include an identification of the request and the requestor, a description by the requestor of what is needed and why, a discussion by Information Systems of the need and proposed ways to meet it, the action recommended by Information Systems supported by reasons and cost estimates, and all approvals necessary to authorize proceeding with what is requested.

TABLE 2-2 Parts of a Service Request

- Identification
  - Date Initiated
  - Requestor Name and Organization
  - User Management Signature
  - Service Request Title
  - Date Received
  - Identification Number
- Need for Service
  - Problem or Area To Be Improved (Not Description of Solution)
  - Savings/Value of Improvement
- Discussion of Need and Proposed Solutions (usually separate document) by business systems analyst
- Recommendation of Information Systems
  - What Should Be Done
  - Preliminary Estimates (Calendar Time, People and Equipment Time, and Cost)
  - Approving Signatures
- Authorization To Proceed
  - Information Systems Planning Committee Signatures
  - Information Services Signatures
  - User Management Signatures

INITIATING THE SERVICE REQUEST

Service requests are usually required for the services of Information Systems. Users normally initiate them; although members of Information Systems may also do so. They are not necessary for work users will do themselves, using personal computers or an information center.

In initiating a service request the user describes what is needed, why it is needed, the value of doing what is requested, and when it is needed.

The user should identify himself by name and phone number, identify his organization, and give the date he initiated the service request.

If he is comfortable doing so, the user may give a title to the service request. Otherwise, he may leave this for Information Services to enter, since titles should be assigned in the light of other requests that have been made.

ID ______________

INFORMATION SYSTEMS SERVICE REQUEST

INITIATION

Title: ____

Initiated by: ______________ Name: ______________________ Date: __________
Organization: ______________________________________ Phone: __________
Request approved by: _______________________________ Date: __________
What is needed: ______________________________________
____
____
____
____
____

Reason for the request: ______________________________
____
____
____

Savings/Value of improvement: ________________________
____
____
____

Date needed: ____

RECOMMENDATION

Action: ____
____
____
____
____

Reasons: ____
____
____
____
____
____
____

Preliminary estimates: _______________________________
____
____
____
____

APPROVALS

Information Systems: ________________________________ Date: __________
Service Request Review Committee: ___________________ Date: __________
Requestor: ____ Date: __________

FIGURE 2-2 *Sample service request form*

The user should describe his need. This may be something that is a problem to him or an area where he believes improvements should be made. He should not attempt an exhaustive discussion of what he wants, since Information Systems will work with him to develop a further description of what is needed. But attaching background material is acceptable and helpful.

Two things are particularly important in this description:

1. He should only describe the problem—what should be done—not how to do it.
2. He should identify the basic functional area and the data that may be affected.

For example, he should say something like:

The Personnel Information System should be modified to allow the inclusion and reporting of safety information on employees.

This request is specific about what he needs and what area is affected.

He shouldn't say something like:

Add a fifteen character field to the Personnel Database after employee number.

This request is too specific; it gets into the "how" of satisfying it. Even the most experienced Information Systems analysts may sometimes be drawn into pursuing the user's choice of a solution rather than studying the actual problem.

Nor, should he say:

OSHA has established a requirement for maintaining information that we do not not maintain. Please correct this.

This request is too vague. It does not specify the type (personnel) of information that is affected. When someone in Information Systems reviews the request he may not know what area is affected, until he calls the requestor.

He should give the reasons why he is making the request. In the example of the change to the Personnel system, he might say something like:

OSHA has established new guidelines for maintaining information that we do not currently have.

If the problem is that two financial reports cannot easily be reconciled, he might give as a reason:

> *Correcting the problem will eliminate confusion between the reports and manual work required to reconcile them.*

The user should also try to quantify the benefit of solving the problem he is describing. He need not do a thorough study of its economic advantages, but should make specific statements of things he can estimate. Examples are:

> *Not providing this information will expose the company to penalties of $1000 a day beginning August 1.*

> *Correcting the problem will save one man-week of effort a month by an associate accountant.*

He should also obtain the approval and concurrence of his management. Doing so gives him an opportunity to discuss with his management what he wants and why he wants it. These discussions help to produce requests that better fit the plans of user management. They may even lead to the request being changed or cancelled.

After the user has prepared the service request and his management has reviewed and possibly modified it, the user should send it to Information Systems.

REVIEWING THE SERVICE REQUEST

Information Systems' review of a service request involves several steps. First, a member of the Information Systems clerical staff will log the request, assign an identification code to it, and acknowledge its receipt. The service request should then go to the manager responsible for the business systems of that part of the organization to which the requesting user belongs. In a small organization one systems development manager will be responsible for all business systems analysis as well as systems development. In medium to large organizations different managers may be responsible for different areas of the organization, but each will be responsible for both business systems analysis and systems development. In the largest organizations one group of managers will be responsible for business systems analysis and another group for the systems development functions for the different parts of the organization.

TABLE 2-3 Review of Service Request

WHO	WHAT
• Clerical	• Logs • Routes
• Management	• Reviews • Determines next step • Assigns
• Business Systems Analyst	• Studies • Develops problem statement • Evaluates alternatives • Determines feasibility • Recommends resolution
• Management & Planning Committee	• Resolves • Establishes priority

The manager who reviews the service request should consider such different issues as the organization's data strategy, functional systems plans, other requests from the same area of the organization, existing or planned development projects, and the feasibility of what is requested on its face alone. Based on this review the manager may recommend what action should be taken without any further study being made.

If further study is required, he will assign a business systems analyst to develop a problem statement for it. Many organizations still have only one basic job classification for systems analysts. It includes both the study and analysis functions and the development functions.

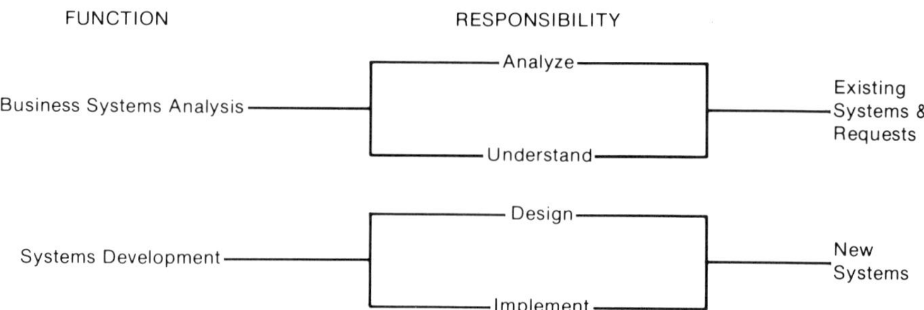

FIGURE 2-3 Parts of systems analysis function

Other organizations have split the functions into two different job classifications. Figure 2-3 shows the two distinct functions that make up what we think of as systems analysis. A business systems analyst should be responsible for this part of the work, if the organization has both business systems analysts and system development analysts.

DEVELOPING THE PROBLEM STATEMENT

The problem statement is a description of what the user wants done, why he wants it done, and when he wants it done. Writing it is the first part of the work that a business systems analyst will do on a service request. It elaborates on what the user has already written. If the user has been thorough and has described the problem instead of its solution, no additional problem statement may be necessary.

Although the problem statement is usually too long to be written on the service request itself, it should be kept in a file that is established for the service request. It is a part of it, even though it may not be physically attached to it.

In order to develop a problem statement the business systems analyst should have some knowledge of business systems and the user's functions. He should also be skilled in interviewing and documenting.

The business systems analyst must answer the kinds of questions shown in Table 2-4 to learn what the user wants.

TABLE 2-4 Questions To Learn What User Wants

REGARDING WORK TO BE DONE OR INFORMATION REQUIRED

- What is to be produced
- Contents
- Purpose
- When it should be done
- Criteria for its production
- Its media
- Disposition
- Frequency of production
- Volume
- Response time after stimulus to produce it

TABLE 2-4 (Continued)

REGARDING INFORMATION TO BE USED

- What information is to be used
- Whether it exists in machine-readable form
- Volume
- Updating frequency
- Updating volume

REGARDING SOURCES OF INFORMATION

- Input that must be processed
- Media
- Volume
- Frequency of processing

Work To Be Done or Information Needed

What is needed and contents The analyst should describe what the user wants and the information elements it should contain. For example, the user may be the manager of an airline marketing organization who wants information reported on frequent flyers. The analyst should interview the user to learn what information he wants reported, such as passenger name and address, along with the date, flight number, and miles flown for each flight.

Purpose The analyst should learn why the user made his request. Doing so may cause him to rewrite what the user is requesting. For example, the user may want a report on frequent flyers to control the awarding of prizes based on how many miles a passenger has flown with the airline. After learning that, the analyst may focus the problem statement on a system for controlling frequent flyer contacts, instead of just producing a report.

When should the work be done Regardless of when the user may want a report produced—for example, every month, or every day, or only once—the analyst needs to learn when the user first wants to see it.

Criteria for production The system should produce reports or do work only when certain criteria are met. The criteria for producing a frequent flyers report might be that two weeks have elapsed since the last one was done.

Media and disposition What media should the system produce information in? Should it produce only paper, or a screen, or should it produce both, or some other form of output. In the frequent flyer example, should the system print both a notice for the passenger as well as an internal control report?

Frequency and volume The analyst also needs to learn how often the work should be done or the information reported and how much information will be produced each time.

Response time Required response time is the amount of time that may elapse between the occurrence of a stimulus and the response to it. If the beginning of the week is the "stimulus" for reporting the information, how soon after the week begins should it be reported? Response time is more meaningful for producing screens or reports in response to inquiries. If the stimulus is an unscheduled request for information on frequent flyers, how soon after the request arrives should the system produce the information?

Information To Be Used

Information to be used The analyst must identify the information needed to satisfy the service request. It may already exist in a file or database. But, if the requesting user organization does not have much experience in using automated systems, the user may not know about existing data. Or the data may be in a paper file or may exist in a collection of completed forms, such as ticket copies.

Media The analyst should identify the media of the existing information. The analyst should avoid the natural inclination to begin designing at this point. If the service request is for a frequent flyer system, for example, and the information to be used exists only in the form of ticket copies, it is easy to visualize updating a database that will be used for the reporting. The analyst should try not to do this. He should only focus on the ticket copies as the source of the information that is needed. Determining the ultimate form or medium of the ticket information will be done when the new system is designed.

Volume and frequency of updating Estimates of how much information will be kept and how often it should be updated are valuable in evaluating the request.

Volume to be updated An estimate of how much information will be updated during each updating period, indicated by the frequency of updating, is also helpful.

Sources of the Information

Source input to be processed The source input is that input that the system processes. The system may validate it, summarize it, modify it, and update the system's files, identified above as "information to be used," from it. The system's files may be as different as a machine-readable database or a file of paper forms. At this point, while avoiding slipping into physically designing the new system, the analyst should identify the sources of the information that will be used. If an existing database will be used, the analyst may not need to identify the known sources of information. If a file of forms will be used, the analyst should also identify the sources of the information that goes on the forms. For example, if the file is of approved purchase orders, the sources may include catalogs, price lists, and requisitions.

Input media and volume The analyst should also document the media of the source input and estimate the volume that would be processed during an updating cycle.

Frequency of input processing The analyst should specify how often the input should be processed. This represents the timing of the updating cycle. If the information from the system is required weekly, the input probably should be processed at least weekly, so that the latest data will be used.

AGREEING ON PROBLEM STATEMENT

The user and Information Systems should agree on the problem statement before additional work on the service request proceeds. The analyst should consider the first version of the problem statement that he writes as only a draft. It will not be complete until he presents it to the user, secures the user's comments, and makes any modifications which are necessary to make it a document with which the user will agree.

Individual users sometimes may seem to be particularly difficult in agreeing with a problem statement. But the analyst must persevere to produce one with which the user does agree, since the problem statement is nothing more than an elaboration of what the user wants.

If the user has requested a specific solution to his problem, such as a frequent flyer report, and the analyst has restated his request to be a more general description of his needs, the user may not agree at first. But, most users will concur when the advantages of a having a thorough review of their needs by Information Systems professionals are explained to them.

CONSIDERING ALTERNATIVES

The understanding gained through developing the problem statement should aid the business systems analyst in developing alternate solutions for the problem. Remember the problem statement is not a resolution of the service request; it is simply a thorough description of the problem raised by the originator of the service request.

Listing Alternatives

Taking the time to list alternatives ensures giving careful consideration to the problem. The alternatives are a list of possible solutions to the problem described in the problem statement and a list of possible approaches or methods for accomplishing them. They include the kinds of things shown in Table 2-5.

The business systems analyst responsible for the resolution of the service request should be responsible for developing the alternatives, but the analyst should receive the assistance of one or more design analysts. In order for them to contribute viable alternate solutions, the analyst will first need to explain the problem to them.

The time spent on developing the list of alternatives should be commensurate with the probable size of any project undertaken to resolve the service request. If the probable size of the project is less than one manweek, one-half hour of careful consideration of alternatives should be adequate. If the project may require between a manweek and three manmonths, spending several hours developing alternatives should be appropriate. If the project may require more than three manmonths, one or more mandays should be spent generating alternatives.

TABLE 2-5 Kinds of Alternate Solutions

SOLUTIONS	METHODS
• Non-computerized solution	• Mainframe or micro implementation
• Customized system or system modification	• User developed
• Modified package	• Information center developed
• Unmodified package	• Information Systems developed
• Service bureau implementation	• Third party developed
• Utility	

The list of alternatives should be practical and should fit both the nature and scope of the problem. The analyst should also remember he may not be able to find a viable solution for every problem. If the problem costs the organization $500 per month and the cheapest possible solution would obviously cost at least a hundred times $500, then there is no viable solution to the problem. The analyst should not consider the costs of using different alternatives at this point, but sometimes, such as in the example cited, the differences between the scopes of the problems and their possible solutions will be obviously large.

In fact, human nature makes declaring that a problem has no viable solution a difficult thing to do. Users often expect automation at any price. Analysts and Information Systems managers presume that they can solve any problem with a system of some sort. The analyst and his managers should discipline themselves to presume that they may not be able to solve every problem.

To develop a useful list of alternatives the analyst must be able to learn what services different components of the organization and outside agencies provide, and what kinds of hardware, communications capabilities, systems, databases, packages, and utilities are available.

Describing Alternatives

Developing and analyzing alternatives requires independent thinking. The analyst should try to be creative in developing alternatives. He should be wary of the common pitfalls, such as:

- Analysts gravitate to what they know.
- Persons who work with mainframes devise mainframe solutions.
- Software developers want to build not buy.

Table 2-6 describes what the analyst should include in a description of the alternatives proposed for a service request. The description of each alternative must be sufficient for determining what effect the availability of systems professionals will have on it, how it relates to the organization's data strategy and its functional systems plan, and the estimated costs of developing and operating it.

The analyst should describe the scope of each alternative. The scope tells which parts of the problem statement each alternative will or will not satisfy.

The analyst should identify who will develop the system done in response to the service request. The alternative may be based on the assumption that Information Systems will do the work, or an outside

TABLE 2-6 What Descriptions of Alternatives Must Contain

- Scope
- Who Will Develop
- Development Approach
- Communication with Other Systems
- Relationship of Databases
- Processing Peculiarities
- Additional Hardware, Software, Facilities, and Services Required

organization may be used, or the group originating the request may do at least part of it.

The analyst should describe how the development work will be done, including whether prototyping will be used. "The Application of Prototyping" in Chapter 1 describes when prototyping should be used. The analyst should also highlight any uses of such things as fourth generation languages for the production version of the resultant system, since such things may affect the processing cost estimates for the alternative.

The analyst should describe how the resulting system will communicate with existing or planned systems. If the system will produce new inputs for or will require a new input from existing systems, the work to be done in the existing system should be part of this alternative.

The analyst should describe any relationships between databases the alternative approach will produce and maintain and databases that exist or are planned elsewhere. Identifying redundancies is particularly important.

The analyst should highlight any particular alternative that may require either excessive or scant resources for processing.

The analyst should identify any additional hardware, software, facilities, or services that will have to be acquired for the alternative.

Let us say as an example that the problem to be addressed is the production of a Personnel or Human Resources Information System, including personnel profiles, job classifications and associated salary grades, skills inventory, salary and performance review management, employee education, and wage and benefit plan administration. The system will process the records of approximately 500 salaried and 1200 hourly-paid employees; all of them are currently covered by the mainframe payroll package that the organization purchased and modified two years ago.

The basic list of alternatives includes:

1. Purchasing and modifying the sister personnel package sold by the vendor of the payroll package.
2. Purchasing services from a service bureau using a system that is compatible with the organization's payroll system.
3. Developing a mainframe personnel system.
4. Modifying the payroll system to add additional personnel information.
5. Purchasing two integrated microcomputer packages that perform salary and performance review and wage and benefit plan administration, then developing additional microcomputer systems for the remaining functions.
6. Developing some microcomputer subsystems to aid what would mainly remain a manual process.

All of the alternatives, except for the second, include helping some Human Resources department people to develop their own reports and inquiries through the organization's Information Center.

As an example, the analyst's description of one of the alternatives—that of developing a mainframe human resources system—may indicate it will meet all the requirements of the problem statement; Information Systems professionals will develop and operate it; they will use prototyping to develop it; it will contain a database that will partially duplicate information in the payroll system; the database will be maintained through on-line transactions entered by personnel department people and through extracts of payroll system batch transactions; and, the organization will not need to acquire additional hardware, software, facilities, or services for it.

EVALUATING ALTERNATIVES

Table 2-7 lists the criteria for evaluating the alternatives developed for the service request.

TABLE 2-7 Criteria for Evaluating Alternatives

- Availability of Systems Professionals
- Data Strategy
- Functional Systems Plan
- Preliminary Cost Estimate

TABLE 2-8 Quantifying Need for System Professionals

- Job Classification
- Experience
- Number of Persons
- Beginning Date
- Ending Date
- Mandays

Availability of Systems Professionals

The analyst should list the information shown in Table 2-8 for each alternative to quantify what it requires from systems professionals. This will include the number of each type of professional, their experience, the time they will be needed, and the effort they will need to expend, measured in some multiple of mandays. For example, the analyst might state that doing the alternative will require three months. During that period one senior analyst will be required for 35 mandays (half time), one programmer-analyst will be required for 70 mandays, and another programmer will be needed for 25 days beginning one month after the work commences.

Considering Data Strategy

An organization's data strategy describes what data is important to it, who will create and maintain it, how it will be maintained and communicated, and who may use it.

If an alternative conflicts with the data strategy, Information Systems may eliminate if from further consideration or may change it to conform to the strategy.

In the example used earlier, developing a mainframe personnel system that partially duplicates data maintained by the payroll system may conflict with a norm that essential data, which includes personnel information in this case, be maintained in only one database. The analyst may have to modify the alternative to describe providing access to the data in the payroll database and developing a way to update it, including modifying the responsibility for maintaining it.

Considering Functional Systems Plan

If an alternative violates the organization's functional systems plan, Information Systems may also eliminate it from further consideration or may change it to conform to it.

In the example described earlier one alternative was to modifiy the payroll system to process personnel data. But if the organization's functional systems plan specifies a separate system for Human Resources to assist them in performing their functions, then Information Systems would probably rule out the modification to the payroll system because it violates the plan.

Preparing Preliminary Estimate

Those who will decide the final resolution of the service request need preliminary cost estimates for all the alternatives they will consider. Each cost estimate should include the costs of both developing and operating the system that will be produced.

For an estimate to be realistic, it should be based on such things as the number of reports and screens the system will contain, the structure and contents of the database, the DBMS that will be used, the number of program modules to be written, the programming languages to be used, and the skill levels of the people who will do the work. But, the analyst will not have determined these things yet. He can only make educated guesses of them. The guesses will depend on the detail of the problem statement.

The analyst will base the cost estimate on the guesses made, his experience, assistance from his colleagues and managers, and any factors used in Information Systems for making such estimates. Paradoxically, although the cost estimate will perforce be preliminary, it will be important in the resolution of the service request. And, the original budget for any project based on the service request will be grounded in it.

DETERMINING FEASIBILITY

The feasibility recommendation should state whether the benefits of the requested work outweigh the costs of doing the work.

The feasibility determination differs depending on whether development costs are charged to users, either actually or in management reports. Its flavor also differs depending on whether Information Systems makes the determination alone, or in conjunction with the requesting user, or whether a separate Information Systems planning committee, independent of Information Systems, makes the determination. Regardless of who makes the final determination, Information Systems contributes importantly with its recommendation.

If development costs are not charged to users and if Information Systems is working off a large backlog of requests, a proposed project may have to show a high benefits payoff, possibly as high as 10 times cost, to be considered feasible. This is true whether Information Systems makes the feasibility determination alone or recommends to a committee which makes the final determination.

When development costs are charged to users, then a much lower multiple, possibly as low as an even payoff of costs by benefits, may still allow a project to be considered feasible. If development costs are charged to the user and the user participates in the feasibility determination, little notice may be taken of the feasibility of a high priority request. If development costs are charged to users and an Information Systems planning committee makes the final determination, the committee may consider the actual multiple of benefits over costs much less important than the budget for development costs that the requesting organization has and how the request was evaluated in light of the availability of systems professionals, the organization's data strategy and its functional systems plan.

RECOMMENDING RESOLUTION OF SERVICE REQUEST

By this point in the review of a service request, the Information Systems analyst will have reviewed the alternatives that were developed. If one or more alternatives fit the organization's data strategy and functional systems plan, if the appropriate systems professionals will be available within a reasonable period of time, and if the request was determined to be feasible, then the analyst should recommend the alternative to be followed as the resolution of the service request. Otherwise, he may recommend no action, incorporating the request in other pending requests or as a modification to a project that is in process, or doing a more detailed study of the needs described in the request.

Information Systems management, any Information Systems planning committee, and user management must concur in the recommended resolution of the service request.

ESTABLISHING PRIORITIES

If the resolution of the service request that is approved calls for positive action, the next step is establishing the priority of the project that the service request authorized. The priority of the project will be relative to the priority of the other open and planned projects. Information Systems management, along with the Information Systems planning committee and user management, will determine the priority of the project and possibly adjust the priorities of other projects because of it.

Information Systems will then enter the project into its overall systems development plan for the organization.

PLANNING NEXT TWO PHASES

The final step Information Systems should take is the detailed planning of the first two phases of the project, which are the study of the present system and defining the prototype of the new system. Chapters 3 and 4 describe the work to be done and who in the organization should do it. Those chapters should assist those responsible for developing this detailed plan.

SUMMARY

1. Service requests are formal requests for the assistance of Information Systems.
2. Uses for service requests:
   - Giving visibility to Information Systems' work
   - Ensuring user management reviews them
   - Ensuring user needs are understood
   - Evaluating them
   - Estimating time and costs
   - Determining their feasibility
   - Securing approval for them
   - Assigning them a priority
   - Controlling (not losing) them
3. Parts of a service request:
   - Identification
   - Description of need for service
   - Discussion of need and proposed solutions
   - Recommendation of Information Systems
   - Authorization to proceed

4. The problem statement is a description of what the user wants done, why he wants it done, and when he wants it done.
5. Questions an analyst should ask to develop a problem statement:
   - What work does the user want done or what information does he need?
   - What information is to be used?
   - What are the sources of the information?
6. The alternatives are a list of possible solutions to the problem described in the problem statement and a list of possible approaches or methods for accomplishing them.
7. Criteria for evaluating alternatives:
   - Availability of Systems Professionals
   - Data Strategy
   - Functional Systems Plan
   - Preliminary Cost Estimate
8. The feasibility recommendation states whether the benefits of the requested work outweigh the costs of doing it.
9. The resolution of a service request is the recommendation of an analyst with the concurrence of his and the user management.
10. Establishing its priority is the next step for any project approved on a service request.

Chapter Three

STUDYING THE PRESENT SYSTEM

PURPOSE

You should study the present system to learn what functions it performs, what information it maintains in files to perform its functions, and how it communicates internally and with those systems that provide its input and use its output.

Study of the present system should also disclose areas for improvement, such as those listed in Table 3-1. A new system should improve the handling of large volumes of data, reduce delays, eliminate bottlenecks and redundant processing, and enhance the accuracy and speed of mathematical and logical processes.

The knowledge of the present system's functions provides a checklist of possible requirements for the new system. The word "possible" is used since the new system will do things in an improved manner and will not simply be a speeded-up version of the present system. Knowing the present system's inputs and outputs provides minimum communication requirements for the new system.

The information maintained in files by the present system is an excellent beginning for what must be maintained in a database for the

TABLE 3-1 Present System Areas for Improvement

- Volume
- Bottlenecks
- Delays
- Redundancy
- Mathematical Processes
- Logical Processes

new system. The present system may maintain more information than is needed for its functions and some of it may be redundant, but everything necessary for its functions is there at least once.

METHOD FOR STUDYING THE PRESENT SYSTEM

Table 3-2 lists the approach recommended here for studying the present system. Table 3-3 lists the four basic tools used to study existing systems. All are valuable. This book recommends the use of all of them. The essential question when an analyst is using them, though, is what is he following? Is he following the processing or the data?

Following the processing is the common approach. Yet it is the most difficult and the one requiring the most skills on the part of the analyst. The linchpin of the approach is the interview. The analyst must be skilled at interviewing. As he learns about the processing through interviews, he must be able to recognize whether he is receiving a complete description of the processing. This need plus having to focus the interviews on new areas require that he also have a good knowledge of the organization's business and operations. This approach requires a talented analyst. Sometimes analysts are unable to shake their biases and predispositions, causing them to direct the interviews towards a solution they already favor. The questions then become leading questions.

TABLE 3-2 Approach for Studying Present System

- Collect information on present system with document analysis
- Develop schematic diagram of information flow
- Verify it with users in interviews
- Expand information on processing with interviews
- Expand knowledge of processing with observations
- Describe processing in data flow diagram
- Review it with users
- Organize study materials and notes into project file

TABLE 3-3 Tools for Studying an Existing System

- Interviews
- Questionnaires
- Direct Observation
- Review of Documents

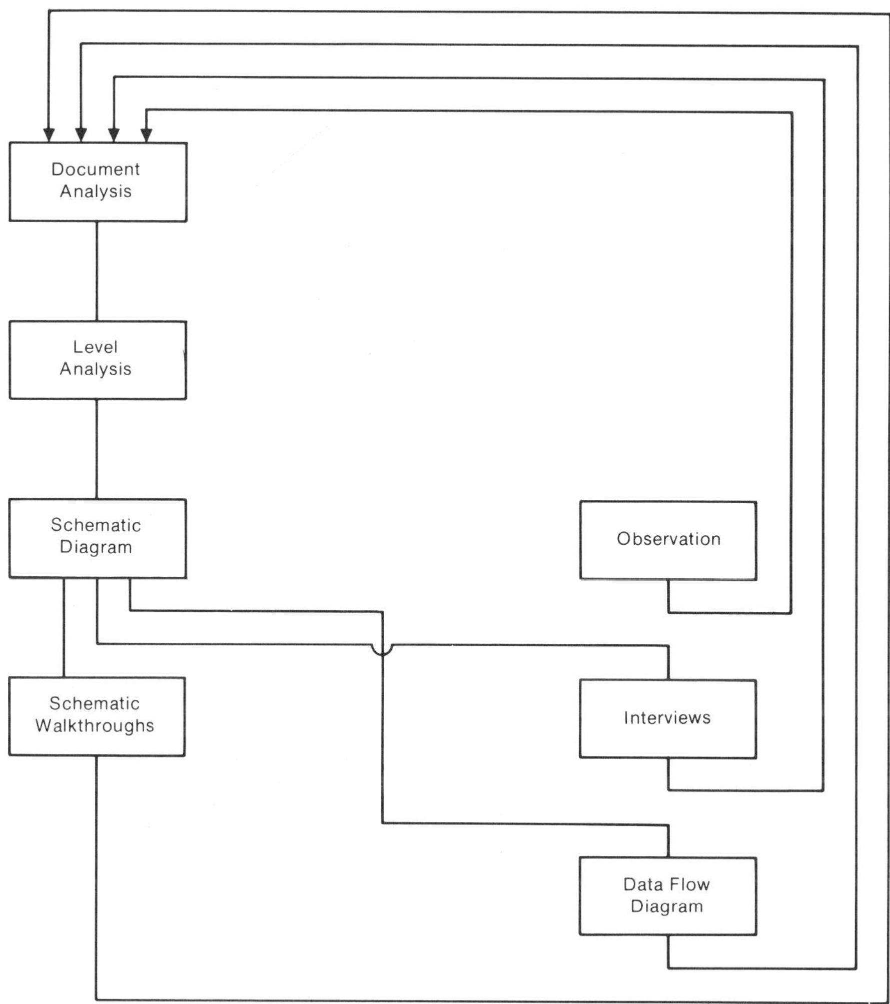

FIGURE 3-1 Steps in studying present system

The method this book recommends is to follow the flow of information as it is recorded on documents. Figure 3-1 depicts the approach schematically. Even though interviews, questionnaires, and observation are a part of the approach, the techniques for following the document information flow provide mechanical aids for ensuring that all the processing in the area has been discovered.

A document is anything used to record, modify, transmit, or file information. A document is anything information is written on or printed on; it's anything on which quantities are extended; it's a list of item numbers posted from one source that have unit prices added to them from another source; it's a telephone call in which a customer is notified of the items on an order that have been placed on backorder; it's the individual file folders maintained for each customer and the various pieces of paper containing information that are kept in the file folders.

The "present systems" that were studied twenty years ago, or even ten years ago, were usually manual and office machine processes. Today computers are everywhere. Even small companies use computer systems which may run on microcomputers or service bureaus. The present system today is usually a combination of manual, office machine, and computer processes. A file may be a computer file; a document may be a report printed on a computer. But, the method of following the information flow on documents remains, regardless of the medium used to prepare or to store the document.

Products of the Study

Table 3-4 lists what the study of the present system will produce. The schematic diagram depicts the information flow, timing, and volumes in chart form. The document description worksheets are the source for it.

Each document description worksheet describes one document. It should also have as attachments a sample copy of the document and of each of its sources. The samples should contain actual data.

A logical data flow diagram depicts the functions the present system performs, and the main files it uses and maintains. The diagram summarizes the information discovered in the document analysis and schematic diagram preparation processes, in the observation of the operations, and in the walkthroughs of the schematic diagrams.

The notes are the analyst's notes organized by information element, by operation on the data flow diagram, and by document identification. Some notes pertain to the meaning, use, processing, validation, or relationships of information elements. Some analysts

TABLE 3-4 Products of Present System Study

- Schematic Diagram
- Document Description Worksheets
- Data Flow Diagram
- Notes

view collecting them as beginning a data dictionary. You should organize them by element name. Other notes expand on the processing the data flow diagram depicts. You should annotate them by operation name. Another set of notes describes the documents which compose the present processing. You should relate them to the document identification.

Keep the notes together in one set of files. You should be able to retrieve information from them with a combined alphanumeric index of element name, operation name, and document identification or name.

DETERMINING SCOPE AND SIZE OF STUDY

What area will you study? How deeply will you study it? What methods will you use in the study? What the new system will do and the areas of the organization that it will serve should point you to where to begin the study.

If, for example, the new system will be a Human Resources information system, including personnel profiles, job classifications, salary grades, skills inventory, salary and performance review management, employee education, and wage and benefit plan administration, as described in Chapter 2, then the study should include the groups within Human Resources that perform those functions.

As the study proceeds you may find several inputs coming from an area or several outputs going to an area that you had not planned to study. In the case of the Human Resources system, several outputs go to the Payroll system area. That may lead you to study how those outputs are used in Payroll. Having that information should improve the design of the new system for the interface with Payroll.

If the new system and the project to develop it will be large, then the study should be a full one, using document analysis techniques along with walkthroughs and interviews. If the project will not be large, the study should at least consist of interviewing people to discover the information flow, collecting sample documents containing actual data, and preparing brief schematic diagrams which are reviewed with user people.

If the study will be large and will cover the work of more than one group, you may want to do the document analysis and draw the schematic diagram separately for each group. You can show the communication between the groups with connectors on the schematic diagram. You can later do one data flow diagram to depict the processing done by all the groups that you studied.

PERFORMING DOCUMENT ANALYSIS

Document Description sheets are basically questionnaires used to gather the information needed for the document analysis. Table 3-5 lists the three different worksheets.

The Document Description sheet describes the processing done on the document to be analyzed, how many documents are done, how often they are done, and how they are filed.

The Copy Distribution sheet describes where and how copies of the document are sent.

The Source Document sheet describes the sources used in working on the document.

The steps to be followed in doing the document analysis are shown in Table 3-6.

TABLE 3-5 Document Description Sheets

- Document Description Sheet
- Copy Distribution Sheet
- Source Document Sheet

TABLE 3-6 Steps in Performing Document Analysis

- Gathering Information on Documents
- Collecting Sample Documents
- Organizing and Checking Information Gathered

Document Analysis Approach

In determining the scope of the study, identify the area of the organization whose work is to be studied. You should study all the manually or office machine-prepared documents that are prepared, modified, or verified within that area. Table 3-7 describes the categories of documents.

Both modified and verified documents come from outside the area being studied. Work is done on them in the area. Sometimes documents come from outside the area and are used only as sources, but they are also filed. To show this on the schematic diagram, you must either define the file as a document or show the source document as

TABLE 3-7 Document Categories

- Prepared—originated in the area that is being studied
- Modified—information is added to a document that was originated outside the area that is being studied
- Verified—information that was added outside the area that is being studied is checked inside the area

being modified within the area. In this case the "modification" would be the filing of the document.

You must apply judgement to decide whether to study documents prepared by computer systems that are considered part of the area that is being studied. If, for example, you are studying the Payroll area which uses a large Payroll system that runs on the organization's mainframe computer, should you analyze the documents produced by that system? Recall that the purpose of studying the present system is to learn what it does, what information it maintains, and how it communicates with other systems. If the documentation of the computerized system provides this information, you do not need to redo it into a document analysis format. Instead, you may show reports coming from it as source documents and information going to it as document copies that are distributed.

If the documentation of the computerized system, whether it runs on a mainframe, minicomputer, or microcomputer, is not adequate, you should describe it using the document analysis approach.

Who should do the analysis depends on the size of the area to be studied along with the number of analysts and the calendar time available for the information gathering.

The source of information for the document analysis are the people and their supervisors in the area to be studied. Analysts can sit with them and fill out the sheets in response to questions they ask about the documents. The analysts should be able to do this relatively quickly, because they understand information flow and are familiar with the sheets.

Another approach is to prepare detailed instructions and to provide training and guidance on completing the sheets so the people in the area can do the sheets themselves. After expending the time to write the instructions, which should be a one-time effort, and training the people, one analyst can guide several people as they prepare document analysis sheets. This approach can lead to gathering a large amount of information in a relatively short time period.

Whether analysts do the analysis themselves or whether people in the area being studied do it, beginning is easy. Start by analyzing one

TABLE 3-8 Document Analysis Control Lists

- Documents by Document Name
- Documents by Document Identification
- Documents Analyzed by Document Identification
- Document Distribution by Organization Sent to
- Source Documents by Sending Organization
- Source Documents by Document Identification

of the originated, modified, or verified documents that a person does and continue until you have analyzed all such documents. If several people perform exactly the same tasks, you need to analyze the work of only one of them. If several people perform similar tasks with some variations, you should analyze all the work of one of them and only the differences in the work of the others.

Regardless of who is completing the sheets, one analyst must control what is being analyzed. To do this he needs to maintain at least the lists shown in Table 3-8. The analyst uses the document lists to relate names to identifications and identifications to names. The analyst also may maintain separate lists of documents analyzed by name and by identification to control whether a document has been analyzed or not. The document distribution list is useful as a checklist of documents that should be found as sources when the work of the area where they are sent is analyzed. Likewise, the source document list identifies documents that should possibly be analyzed if the area where they are produced is studied. Using a DBMS on a microcomputer can aid in maintaining these lists.

Document Description Sheets

Figure 3-2 shows a sample a Document Description sheet form. The following text describes the information that should appear on a completed sheet.

The *System Code* is an arbitrary code for individual system areas as identified in the organization's functional system plan. Examples might be "PAY" for Payroll or "MNC" for Manufacturing Control.

The *Document Identification* is the form number or some other unique identifier of the document within the system. When form numbers do not appear, the analyst may have to assign a number to a document. If the document originated outside the area and is modified or verified, add a "-M" or "-V" suffix to the document identification. That

DOCUMENT DESCRIPTION

GENERAL

System: ______ Doc. Ident.: ______ No. of Cys: ______
Subsidiary: ______ Cost Cntr: ______ Type: Orig: Mod: Verify: ______
Document Name: ______ Ident. Type: ______
Procedural Manual Reference: ______
Is document a standard form used for different purposes? _Yes: ____No: ______

DISTRIBUTION

Sent from: ______
Subsidiary: ______ Cost Cntr: ______ No. of Cys Received: ______

FILING

Is document filed? _Yes: ____No: ____No. in active file: ______ Inactive: ______
Basis for retention: ______

PROCESSING

Frequency: ______ Time Required: ______
Volume Average: ______ Maximum: ______
Method: ______

PURPOSE

Document Purpose: ______

PREPARATION

Document Description Prepared by: ______ Phone: ______Date: ______
Person: ______
Subsidiary: ______ Cost Cntr: ______

FIGURE 3-2 Document description form

will distinguish the modified or verified document from the unmodified or unverified source document which should also appear on the source document sheet.

The *Number of Copies* is the number of copies of the document that are originated, modified, or verified—whatever the work being analyzed is. If five copies of a document are originated, but only two modified in the area being studied, the number should be "2."

If you are studying a large organization composed of multiple subsidiary organizations, use the *Subsidiary Identification* to distinguish in which subsidiary the work is being studied. This is valuable particularly when you study the work of similar areas within more than one subsidiary company.

The *Cost Center* is the additional coding which, along with Subsidiary, will uniquely identify the area of the organization where you are doing the study.

The *Type of Processing* indicates whether the principal work done on the document is originating it, modifying it, or verifying it.

The *Document Name* is either the name that actually appears on the document or the name by which it is commonly known. In the absence of either of them, assign it a name that describes it.

The *Identification Type* refers to the Document Identification described above. It indicates how widely the document identification is known. Table 3-9 gives a set of sample codes that could be used.

If the way to do the work on the document is described in any sort of procedural manual, give the *Procedural Manual Reference* here.

If the document that is being analyzed is a *Standard Form Used for Different Purposes*, indicate that here.

If the document is originated elsewhere and modified or verified in the area being studied, *Sent from* identifies the person who sent it and that person's *Subsidiary* and *Cost Center* Identifications.

The *Number of Copies Received* applies to documents sent from some other area.

TABLE 3-9 Sample Document Identification Codes

EXT	Externally Assigned
CO	Assigned at the Company Level
DIV	Assigned at the Division Level
DEPT	Assigned at the Department Level
DA	Assigned for the Document Analysis
OTH	Other

If one or more copies of the document is filed after the work on it is completed, indicate that for *Document Filed.*

The *Number in Active File* is an estimate of how many documents are kept in the active file. An active file is one in which documents are held to be used as sources for other documents.

The *Number in Inactive File* is an estimate of how many documents are kept in the inactive file. An inactive file is a history file that provides a record of action that was taken.

The *Basis for Retention* indicates what determines how long documents are kept in both the active and inactive files. It may be a length of time, such as two years. It may be an occurrence, such as "until employee terminates."

The *Frequency of Processing* is the frequency with which the document is originated, modified, or verified. Sometimes work may be done on a document with a different frequency than its completion. For example, a person may post to a document daily, but it is not complete until the end of the week. In such a case, the most frequent work would be measured—the frequency would be daily.

The *Time Required To Process* entry is an estimate of the time required to complete one document. Examples are "1 Hour" or "1½ Days."

The *Volume* entries are estimates of the average and maximum number of documents that are originated, modified, or verified during any month. If the period indicated is less than monthly, for example quarterly, make the volume estimates for this less frequent period.

People often use a mixture of methods in working on documents. They may type some information on it and later post other information by hand. The *Processing Method* should indicate the most mechanical means used to work on the document, if more than one means is used. In the example just used, work that is partly manual and partly typewriter, indicate "Typewriter" as the processing method.

The description of the *Document Purpose* or function should indicate what the document is and what the work done with it is.

The *Document Description Sheet Prepared by* identifies the person who completed the Document Description sheet.

Copy Distribution Sheets

Figure 3-3 shows a sample Copy Distribution sheet and the following text describes using it.

The *System Code, Document Identification, Subsidiary Identification,* and *Cost Center* are the same as those for the Document Description sheet. They are used on this sheet to tie the information about copies back to the copied document.

Prepared by: ______
Date: ______

COPY DISTRIBUTION

DOCUMENT ANALYZED:

System: ______ Doc. Ident: ______ Subsidiary: ______ Cost Cntr.: ______

COPY DESCRIPTION		SENT TO					TRANSMITTAL	
Copy Ident.	Copy Description	System	Subsidiary	Cost Cntr.	Cost Center Name	Location	Method	Time

FIGURE 3-3 Copy distribution form

The *Copy Identification* identifies the individual copy of the document. If the document contains a preprinted copy number, use it. Otherwise, assign a number to identify the copy.

The *Copy Description* is text used to identify a copy. If a copy contains a title, such as "Customer Copy," use it. Perhaps each copy is color-coded. If so, use the name of the color. Otherwise, assign a description to the copy to tell how it is used.

Sent to System identifies the system to which a copy is sent.

The *Sent to Subsidiary Identification* identifies the subsidiary to which a copy is sent.

The *Sent to Cost Center* identifies the cost center or department to which a copy is sent.

The *Sent to Cost Center Name* is the name of the cost center or department to which a copy is sent.

The *Sent to Location* identifies the location.

The *Transmittal Method* indicates how the copy is sent to the receiving area. Choices would include company mail, U.S. mail, telephone, and telecopier.

Transmittal Time is the amount of time that elapses between sending a copy of the document and its arrival at the receiving point.

Source Document Sheet

Figure 3-4 shows a sample Source Document sheet and the following text describes using it.

The *System Code*, *Document Identification*, *Subsidiary Identification*, and *Cost Center* for the Document Description sheet are used on this sheet to tie the information about source documents back to the document for which they are sources.

The *Source Document Identification* is the form number or some other unique identifier within the system for the document. When form numbers do not appear on a document, the analyst may have to assign a number to it.

The *Source Document Copy Identification* identifies the individual copy of the document. If the document contains a preprinted copy number, use it. If the source document has been analyzed and the analyst assigned an identification to it, use that. Otherwise, assign a number to identify the copy.

The *Source Document Name* identifies the source document. If it contains a title, such as "Personnel Information Sheet," use it. Otherwise, describe it.

The *Use Type* describes how the source document is used. If it is the source of data, enter "DATA." If it is only used to verify information

Prepared by: ____________________

Date: ____________________

SOURCE DOCUMENT

DOCUMENT ANALYZED:

System: ______________ Doc. Ident.: ______________ Subsidiary: ______________ Cost Cntr: ______________

SOURCE DOCUMENT DESCRIPTION					SOURCE DOCUMENT RECEIVED FROM				
Doc. Ident.	Cy. Ident.	Document Name	Use	Med	System	Subsidiary	Cost Cntr.	Cost Center Name	Location

"Use" = Use Type

"Medium" = Medium of Source

FIGURE 3-4 **Source document form**

on the document that is being analyzed, indicate that. If it is used for both data and verification, indicate both.

Indicate the *Medium* of the source document. Examples are form, report, catalog, phone, and telecopier. If information is provided verbally, whether in person or by phone, it should be shown as a source document, even if you do not know what the person providing the information takes it from.

The *Source System* identifies the system from which the source document comes.

The *Source Subsidiary Identification* indicates the subsidiary of the organization from which the source document came.

The *Source Cost Center* indicates the cost center or department that sent the source document.

The *Source Cost Center Name* is the name of the cost center or department that sent the source document.

The *Source Location* is the location from which the source document was sent.

Document Analysis Sheet File

Prepare a file for each analyzed document containing the document analysis sheets and a completed set of all copies of the analyzed document. The sets should contain actual data. If, because of the nature of the document, for example, a paycheck, you cannot attach a document containing real data, attach a sample with sample data. If it is too difficult to attach a sample set with data, attach a blank document set along with a copy of one containing actual data.

Also, attach a sample of each source document with data on it. If a source document does not contain a form number or a title, write on it the identification which you used on the Source Document sheet. If the source document is too voluminous to attach, for example a catalog, attach a sample page showing the data that is typically taken from it.

LEVEL ANALYSIS

An explanation of the level analysis process is given here to aid you in developing a schematic diagram from the document analysis sheets and the control lists prepared from them. A schematic diagram reads from top to bottom, with end-product documents that are not sources on the top and source documents which themselves have no sources

TABLE 3-10 Materials Needed for Level Analysis

- Paper containing .2 inch by .2 inch squares
- Document Analyzed by Document Identification List
- Source Documents by Document Identification List
- Document Analysis File

on the bottom. The top level is level 1 and the bottom level is the largest number. The level analysis process assists in determining what relative level any document should appear on.

Table 3-10 lists the materials needed to perform a level analysis. The paper should contain at least seven more boxes across the top than the number of documents that were analyzed. The table should also contain at least seven more boxes down than the number of source documents. Figure 3-6 is a sample of an initial level analysis chart.

Follow the steps listed below to perform a level analysis:

1. First prepare the sheet of paper by drawing a horizontal and vertical axis. Label the row across the top next to the horizontal axis as the "LEVEL." Label the second column to the left of the vertical axis as the "USES."
2. List the identifications of the analyzed documents both across the top and down the left side. List the source documents which were not also analyzed documents (P1307 and W4 on Figure 3- 6) below them on the left side.
3. Using the Source Document sheets for each analyzed document, make a diagonal line entry in the square for the intersection of the source and the document whose source it is. For example, on Figure 3-6 an entry is

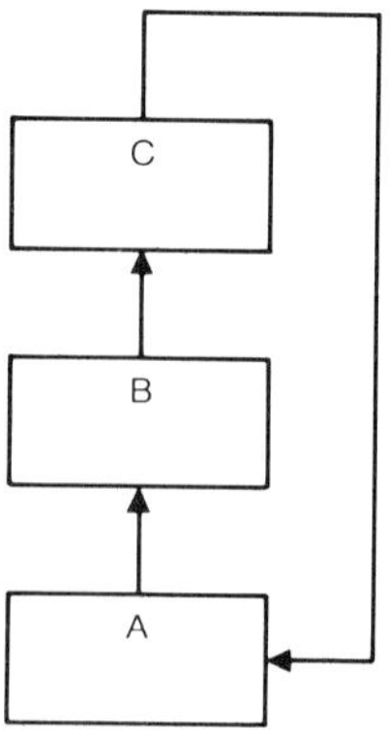

FIGURE 3-5 Schematic diagram of closed loop

	USES	LEVEL	HR14	M2742	PAY60	PO7	PR1375	R3175A
LEVEL			2	4	1	2	3	3
HR14	1	2			X			
M2742	2	4				X	X	
PAY60	0	1						
PO7	1	2			X			
PR1375	2	3			X	X		
R3175A	1	3	X					
P1307		5		X				X
W-4		4	X					X

FIGURE 3-6 Initial level analysis chart

made across from HR14 under PAY60, meaning that HR14 is a source to PAY60.

4. After all the uses of the documents have been noted, scan the number of entries made for each document analyzed and enter zero under the uses for all those that have no uses.
5. Assign a level number of "1" to all documents having zero uses. Post this level number in the box under the document identification across the top of the paper.
6. Go down each column for which a level has been assigned and cross off any diagonal lines that you encounter. This will form an "X."
7. Examine each line across from each document number. Assign the next available level number to any for which you have crossed off all the diagonal lines. In the first case that will be level number 2.
8. Post this level number under the document identification across the top of the paper.
9. Repeat steps 6 through 8 until you have assigned all documents a level number

If after crossing off all the entries you can in step 6, you are not able to assign a level number in step 7, you have either encountered a closed loop or erred.

To find a closed loop, look for a document that has only one use remaining and that has at least one of its uses crossed off for the last set of documents assigned a level. Use the information on the level analysis chart to sketch its part in the schematic diagram. If it is in a closed loop, you will find a situation like that in Figure 3-5 where A is a source to B, B is a source to C, and C is a source to A. The number of intervening levels may be greater than two or three.

If you find a closed loop, assign the next available level number to one of the documents that has only one use not yet checked off. Doing this should enable you to continue with the leveling process. If it does not, it means you have erred or encountered another closed loop.

SCHEMATIC DIAGRAM

The schematic diagram depicts the information flow, volumes, and time in chart form for the information processing being studied. It provides a picture of the document relationships to help you to understand the system. You can also use it to describe and verify what you have discovered about the present system with the persons who work with it and with their management.

A schematic diagram consists of a box for each document, information within the box about the document, triangles depicting outside sources and uses of documents, and lines to show the relationships between documents.

A schematic diagram reads from top to bottom, with end- product documents that are not sources for other documents on the top and source documents which themselves have no sources on the bottom.

Drawing Schematic Diagram

The sources for drawing the schematic diagram are the level analysis chart and the document analysis file. An analyst or documentation assistant can draw it.

The level analysis chart gives you the relative vertical position for placing the symbol for a document. If a document's level number is one, its symbol goes on the top of the chart. If it is two, it goes one level under the top level. If it is the highest value assigned to any document, it goes on the bottom of the chart.

You must determine the relative horizontal position for a document symbol by trial and error. Drawing a rough pencil sketch first helps. Place some of the documents on level one, then draw the documents on level two that are sources to them. Where the difference between copies is meaningful, draw a separate line for the use of each copy. Try to minimize the number of crossovers of lines connecting documents. Continue in this way down the diagram, until you have a good pencil sketch.

Then use the pencil sketch to place the symbols on the final chart.

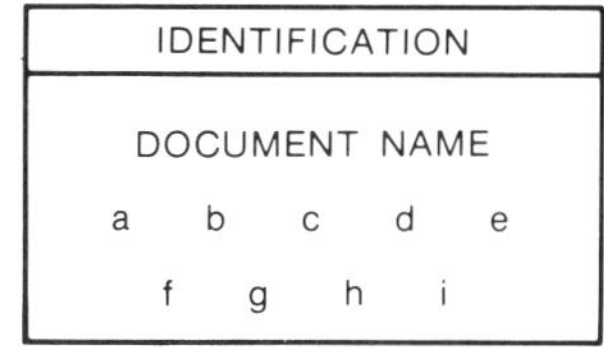

FIGURE 3-7 Schematic diagram symbol format

Another decision you must make is how large the diagram pages should be. Some analysts prefer to draw everything on one large sheet of paper. This approach provides a good overall feeling of the document interrelationships, but it limits the means you can use to reproduce the schematic diagram. Other analysts may draw the diagram on somewhat larger sheets of paper, but for review they reduce it to 8 ½ by 11 inch or 11 by 14 inch sheets.

The symbol for a document and the line connecting a source document or a copy to another symbol should contain additional information in an abbreviated form. Figure 3-7 shows a suggested format for the symbol used for a document and Figure 3-8 shows suggested conventions for drawing the lines connecting symbols. Table 3-11 shows suggested entries for the document symbols.

Figure 3-9 shows a sample schematic diagram. This is for the accounting section that supports the consumer credit insurance line within an insurance company. Consumer credit insurance provides life insurance coverage for consumers for the amount they owe on a consumer debt purchase, such as an automobile or stereo. The accounting section bills the insured for the coverage. When payment is received by the Cash Collection section and it sends a notice of the receipt to this section, it also calculates the sales agent's commission and prepares a check voucher for payment.

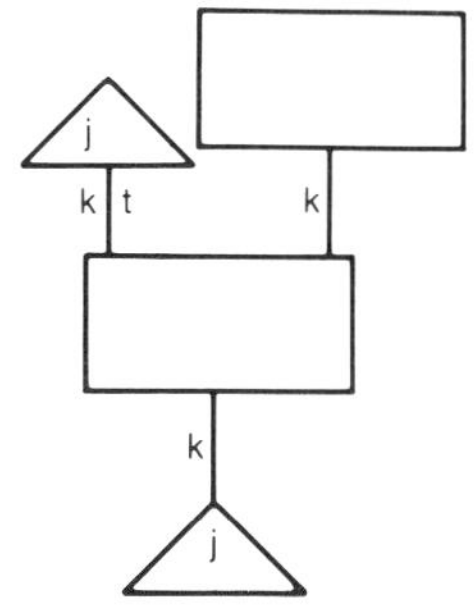

FIGURE 3-8 Schematic diagram connector entries

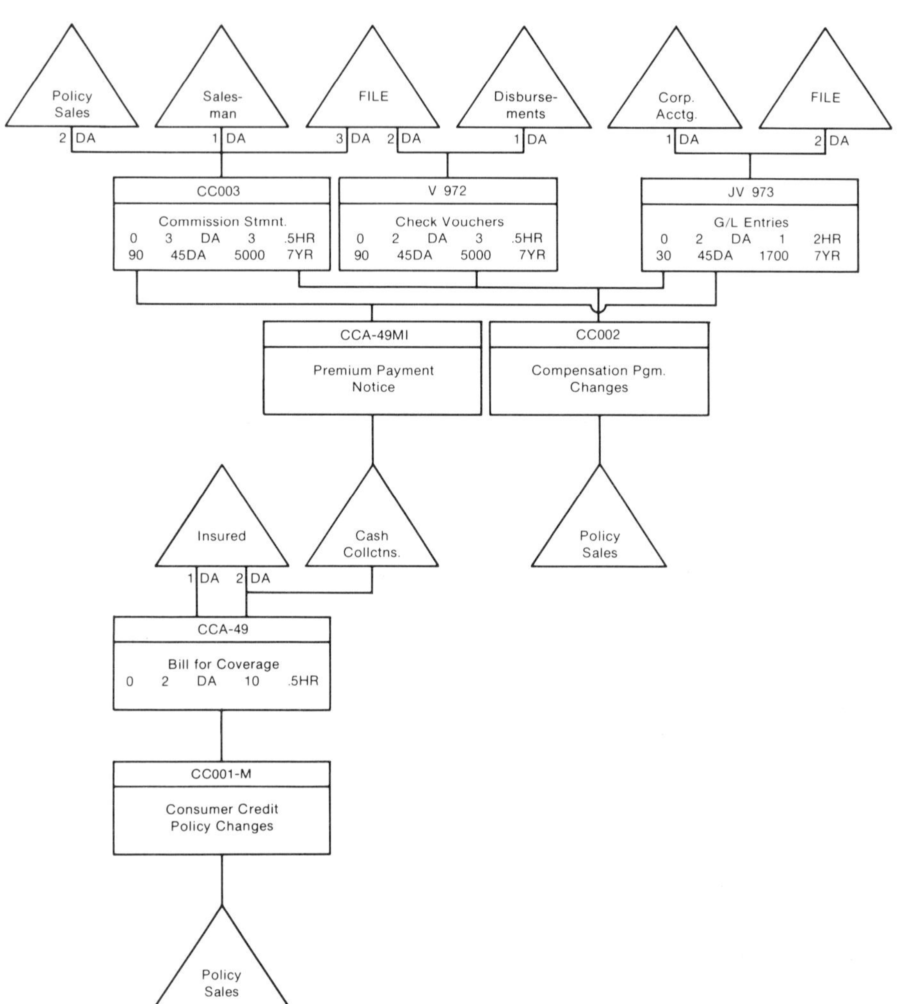

FIGURE 3-9 Sample schematic diagram

The analyst assigned Documentt Identification numbers CC001, CC002, and CC003 to documents that did not contain an identification number. Copy 2 of document CCA-49 is sent to the insured for return with his payment. Cash Collection forwards this to Consumer Credit Insurance Accounting with a notation of the amount of payment received. To show this modified use of a document originally prepared within the department, the analyst assigned it an identification of CCA-49MI. This identification preserves the original organization-assigned number while being a somewhat different number.

TABLE 3-11 Schematic Diagram Symbol Entries

IDENTIFICATION—Document Identification

DOCUMENT NAME—Document Name

- a —Type of Processing: "O" for Originated, "M" for Modified, "V" for Verified
- b —Number of Copies
- c —Frequency of Processing
- d —Volume
- e —Time Required To Process
- f —Number in Active File
- g —Retention in Active File
- h —Number in Inactive File
- i —Retention in Inactive File
- j —Source or Receiving Department
- k —Copy Identification
- t —Transmittal Time

REVIEWING EXISTING PROCEDURES

If written procedures exist that describe the operations in the area of study, review them before completing the first draft of the schematic and beginning the observation of the present system. Even though they are often outdated, they may help you understand what you will see when you observe the actual processing. Although unlikely, they may also disclose an area for study not found during the document analysis.

OBSERVING CURRENT OPERATIONS

Experienced analysts can gain insights into the strengths and weaknesses of an existing system by personally observing it. One or more analysts who understand business functions and are experienced in observing operations should watch and listen to the operations within the area as unobtrusively as possible.

Before beginning the observations, develop a list of functions, using those disclosed by the document analysis completed so far, as a guide to the work done within the area. Add to the list any additional functions that the observation discloses.

You should look for informal processing, bottlenecks, problems, peak times, volumes, and excessive stresses on the system and workers.

Informal processing does not employ documents, at least not in a standard way. Workers may use phone pads or yellow tablets when performing informal processing and throw away the paper they use at the end of the day.

Bottlenecks occur when a large amount of work must be done through a small number of people or equipment. They may even be caused by such situations as too few phone lines or a copier that is too slow.

Problems may be of any kind. Symptoms may be excessive errors, anger or frustration, and the necessity to redo work. You may not be able to identify the cause of the problem, at least at first. It could be inadequate training, incomplete procedures, or heavy-handed supervision.

Estimate the volume of work done during both normal and peak periods. Peak processing periods often create excessive stresses on system and workers; although other factors, such as demands for rapid service, may also cause them.

Notice what workers use and look at while doing their work. Look for informal processes mentioned above that are not done using documents and may not be recorded. Listen to typical phone calls to learn what information comes in them or is dispensed in them. If it's appropriate, actually work for a few days as a member of the group that you are studying.

Observe operations for short periods during which you take a minimum of notes, followed by breaks away from the area, during which you can write more detailed notes. An analyst who writes page after page of notes as he watches what they do makes the average worker nervous.

Ensure before you go into an area that the workers have been told what you will be doing. A straightforward description by user management of the reason for the study usually best allays workers' concerns.

You should observe the operations before you finish the schematic diagram for the area, because you may discover additional functions for study.

You should continue your observations until you feel you have seen all the processing "in action." Complete them as quickly as possible, because performing them will distract the people working in the area.

If the same functions are done in more than one location, you should try to observe at least two of the locations.

The result of observing operations should be additions to the study notes including a description of the bottlenecks that occur in the area's processing, problems that were noticed with the present methods, an identification of the peak periods and the volumes processed during them, other notes on volumes, anything that seems to cause excessive stresses on the system or the workers, and the strengths and weaknesses of the present processing methods.

If the observations disclose additional areas for study, you may need to do document analyses of these areas and add the new information to the schematic diagram.

WALKTHROUGHS

After you have completed your observations of the processing and have drawn the schematic diagram for the area, you should do walkthroughs of the information that you have learned.

Focus the walkthroughs on the schematic diagram, since it presents an overview of the processing within the area. Also, use the document analysis files and your notes. The document analysis files with their attached sample documents are graphic aids for understanding the present processing. Your notes, when organized, will provide additional information on the processing.

The principal business systems analyst responsible for studying the present system should do the walkthroughs. The users who should participate include those workers who have helped with the study, their supervisors of the area studied, their immediate managers, and the user manager responsible for the new system.

The purpose of the walkthroughs differs for the different users. Because of that, you should do them separately. You should do the walkthrough with the workers, supervisors, and managers to verify the accuracy and completeness of the schematic. After you have done this and made any modifications resulting from what you have learned, you should do a walkthrough with the manager responsible for the new system to give him a detailed view of the existing system.

Do the walkthroughs in a conference setting away from the interruptions of the users' business day and with equipment to help in making presentations. This will make the time spent more productive.

To do the verification walkthroughs, you will probably find it works best to start with one of the source documents whose arrival initiates a process and follow the processing trail step-by-step, document-copy-by-document-copy. To describe the processing to requesting management you may first focus attention on one or more of the final outputs of the processing, then mention the individual documents that are part of the processing. In these walkthroughs you will want to point out such things as large files, long retention times, large volumes, redundancies, and transmission delays.

In order to verify what you have learned, the walkthroughs may require a significant amount of time. If you expect to need more than three hours for them, break them into two or more sessions, each about two hours in duration. This should keep them productive while avoiding excessive fatigue for you and the users. You can do the walkthroughs for requesting management on a more general level. Unless the user management wants to delve into the material in detail, one session of two to three hours should be adequate regardless of the size of the area studied.

Do the walkthroughs as soon as possible after the materials for them are ready. Doing the verification walkthroughs may disclose errors and omissions which may lead you to do additional document analysis and to modify the schematic diagram. If this happens, you will want to verify the modifications before doing the walkthroughs with the requesting management.

INTERVIEWS

You should interview users to augment the information about the present processing, its problems, and how the users think it should be improved. In most cases the document analysis, observation of processing, and schematic diagram walkthroughs may have given you sufficient information about the present processing, but you may still need to learn what the users believe the problems are and their thoughts on correcting them.

The interviews may also provide additional information on bottlenecks, problems, peak times, volumes, and excessive stresses on the system and the workers, as the obbservation of the present processing did. And they may give you information about the human factors that make a system work or not work.

Analysts experienced in interviewing techniques and familiar with information processing systems should lead the interviews. The analysts should interview some of the workers, the supervisors responsible for the key areas on the schematic, their managers, the user requesting the system, and the managers of any areas outside the system that provide to it or use from it significant amounts of information.

You may sometimes interview two or three people who are on the same level in the same interview, but you should not interview two or

TABLE 3-12 Interview Checklist

- Features of Present System
- Bottlenecks
- Problems
- Peak Periods
- Volumes
- Excessive Stresses on the System
- Excessive Stresses on the Workers
- Suggestions for Improvement
- Areas Not Studied

more who are at different position levels at the same time. The presence of his manager may inhibit a worker's spontaneity and frankness.

You should begin the interviews as soon as possible after you have completed the schematic diagram walkthroughs. You should also finish them as quickly as possible after beginning them, while still allowing sufficient time to assimilate and document the information that you learn.

Interview the users in their offices or at their desks. Do not invite them to your area nor to a conference room. You can learn more about them and their work habits by observing them in their own environment.

Use a checklist like the one shown in Table 3-12 to ensure you cover all the key points, but try to let the person you are interviewing talk freely. The less he is aware that you are following a checklist, while still feeling you know what you are doing, the more good information he may give you.

Document the interviews in the system notes. If you discover areas that were overlooked, you may need to do a document analysis and schematic diagram followed by a walkthrough of them. You may sometimes believe the study of the present system should be expanded into areas that feed it or use information from it. To do that would probably enlarge the scope of the project. But if it should be done, seek any necessary approvals to do it.

DATA FLOW DIAGRAM

A data flow diagram emphasizes the processing in an area of study by depicting the operations performed, the sources and dispositions of transactions, the files of information, and the transactions that connect them all.

A schematic diagram depicts what is easy to discover—that is the interrelationships of documents. A trained analyst may often presume what the processing is from reviewing a schematic diagram, but the schematic diagram does not explicitly describe it. The data flow diagram emphasizes the processing and deemphasizes the documents.

The basis of the data flow diagram is the information learned from performing the document analysis and preparing the schematic diagram, from observing operations, from doing walkthroughs, and from interviewing.

The analyst responsible for the study should be responsible for the development of the data flow diagram. He may secure the help of documentation assistants to produce the final version of the diagram.

The succinct picture of the present processing provided by a data flow diagram helps you understand the essential elements of the processing by helping you to distinguish them from the present method of doing it. Reviewing it with the requesting user will also help him. The data flow diagram of the present processing is also a good bridge to the definition of the prototype described in the next chapter. It should be the basis of a checklist of the processing that the prototype and the new system must perform.

One objection to doing a data flow diagram is that it is redundant. The schematic diagram depicts the information flow. The information gathered from the schematic diagram, observation of the processing, interviews and walkthroughs is in the study notes. Nevertheless, the extra thinking required to produce the data flow diagram is a valuable addition to the system development process.

Develop the data flow diagram after you have finished the interviews and made any necessary changes to the document analysis and the schematic diagram. Do it while the system is still fresh in the minds of the analysts and the users.

Develop the data flow diagram in sufficient detail to receive the advantages it provides. You will probably begin with a top level diagram which is useful for review with requesting management. Expanding each of the processes in this diagram into a more detailed data flow diagram will increase the diagram's usefulness. Follow a consistent method for drawing a data flow diagram.

Review it with the supervisors of the area being studied to verify its accuracy and completeness. Then review it with the requesting user to provide a succinct overview of the present system. Do the reviews in a conference area away from the users' offices and telephones.

STAFFING INITIAL PROJECT TEAM

Staffing the study of the present system must be done before the study begins. The discussion of it is here, after the description of the work to be done, so you will have a better understanding of the work to be done.

A business systems analyst experienced in interviewing techniques and familiar with the characteristics of information systems should be responsible for the study of the present system.

Some believe it is ideal for the analyst to be familiar with the area being studied. If the Human Resources area is being studied, the analyst should know the Human Resources processes and systems. Others believe detailed knowledge of the system being studied may be a handicap that might cause the analyst not to notice some of the processing. Whether the analyst is familiar with the area or not, he or she should have an open mind and be a good observer of what is *actually* in the present system.

If the area to be studied is large, other business systems analysts may assist the analyst who is in charge. Whether or not the area to be studied is large, documentation assistants may maintain the lists required to control the document analysis, do the level analysis, draw the schematic diagram, draw the data flow diagram, and maintain and organize the study notes.

The management of many organizations, especially large ones, as discussed in Chapter 2, has separated the work of study and analysis from that of design. This separation may provide various economies and advantages. But assigning a person who can stay with the project at least through the definition of the prototype as the analyst responsible for the study will provide valuable continuity for the project. If the project is sufficiently large, this project should be the analyst's only assignment.

PLANNING AND PROGRESS MONITORING

The study of the present system is the first phase of the new project that was initiated by a request. It will set the tone for the remainder of the project. If the work on it is not scheduled and managed well, people will not expect the rest of the project to be scheduled and managed well. Table 3-13 lists some elements that are important for the success of the study.

TABLE 3-13 Essential Elements for Study Success

- Secure interest and cooperation of all levels of users
- Finish with each group as quickly as possible
- Follow a firm schedule allowing for iteration
- Assign one analyst as the leader and support him
- Monitor quality and quantity of work

Secure and keep the interest and cooperation of all the users affected by the study: the people the analysts will work directly with, their supervisors and managers, and the managers who requested the system.

Complete the information gathering portion of the study with each group as quickly as possible. That will minimize the disruption the study causes. Finish the document analysis for a group in about a week. If you study several groups, you may do them independently, but complete each as quickly as possible. Also, do the part of the schematic diagram that applies to the work of each group as quickly as possible, and review it with them when you have finished it. Observe the work of the group as soon as possible after the document analysis. Then follow up speedily with the schematic walkthroughs and interviews. You can do the data flow diagrams and the walkthroughs with requesting management later, because they are not so directly affected as the persons whose work is being studied.

Estimate and plan the work to be done, establish a firm schedule which includes iteration, publish the schedule so the users and Information Systems know what to expect, and adhere to the schedule. Table 3-14 gives an overview of the different tasks to be scheduled.

Estimating what has to be done will be difficult if you are using these techniques for the first time. You can do it. Secure estimates of the number of documents in each area to be studied. Evaluate your confidence in the estimates. Determine whether analysts or users will complete the document analysis sheets. Have those who will do them complete document analysis sheets for a few representative documents. Do a sample level analysis chart. Draft a data flow diagram for an area whose processing you know. Guess how complete and accurate the first information gathered will be and how many new areas will be found requiring new analysis. Imagine what will occur during each phase of the study with each group of people whose work will be studied. Then use your favorite estimating techniques to quantify what you have discovered and imagined, add the numbers and types of persons

TABLE 3-14 Study Tasks To Be Scheduled

TASK	REPEATED FOR EACH GROUP
• Do Document Analysis	Yes
• Do Level Analysis	Yes
• Draft Schematic	Yes
• Complete Schematic	Yes
• Do Schematic Walkthroughs	Yes
• Complete Observation	Yes
• Complete Interviews	Yes
• Draw Data Flow Diagram	No
• Review Data Flow Diagram	No
• Do Additional Document Analysis	Yes
• Do Additional Level Analysis	Yes
• Modify Schematic	Yes
• Do Modified Schematic Walkthroughs	Yes

who will do the study, and complete the schedule and plan. Table 3-14 lists the tasks which will be part of your plan.

Assign the responsibility for the study to an analyst with leadership ability. If the area to be studied is large, provide him sufficient analysts and documentation assistants to complete the study in a timely fashion. Delaying a study until you can properly staff it is better than starting it with too few people. To do that produces an aura of lingering which you do not want the project to have.

The study of the present system will greatly influence the definition of the prototype. Even though this work may seem mundane, carefully monitoring the quality and quantity of the work done by both the Information Systems persons and the users who assist them will ensure a good base for the next phase.

SUMMARY

1. Study the present system to learn:

   - What functions it performs
   - What information it maintains in files
   - How it communicates internally and with other systems

2. Study the present system by following the flow of information on documents.
3. Results of study of present system:

   - Schematic Diagram
   - Document Description Worksheets
   - Data Flow Diagram
   - Notes

4. The schematic diagram depicts the information flow, timing, and volumes in chart form.
5. Document description worksheets are questionnaires to describe the documents that are studied.
6. The logical data flow diagram depicts the functions and the files used by the present system.
7. The notes are the analyst's notes organized by information element, function, and document
8. Steps in studying present system:

   - Do document analysis
   - Do level analysis
   - Complete schematic
   - Do schematic walkthroughs
   - Complete observation
   - Complete interviews
   - Draw data flow diagram

9. Essential elements for success of study phase:

   - Have interest and cooperation of all levels of users
   - Finish with each group as quickly as possible
   - Follow a firm schedule allowing for iteration
   - Assign one analyst as the leader and support him
   - Monitor quality and quantity of work

Chapter Four

DEFINING THE PROTOTYPE

This chapter describes the prototype definition, who should do it, and how it should be done. The chapter first describes what the prototype definition is. Next, it makes suggestions on setting up the project team. Then it explains how to develop the prototype definition. Finally, it tells how to document the definition, to plan the work, and to secure equipment for prototyping.

Why should you do a prototype definition? The motto might be:

Know what you are going to do before you do it.

The prototype definition will be the basis for building the prototype. Since the prototype is the basis for the system, the prototype definition will be the basis for the system itself.

The prototype will be the vehicle for developing the full requirements for the system, and its definition will be the preliminary requirements for the system. Just as you do not need to develop the full requirements before you build the prototype, you also should not attempt to build the prototype without some kind of a definition of it.

Defining the prototype before building it helps users and Information Systems people think through the basic functions of the system. It is a valuable thing to do, even if the prototype may change considerably when the user actually begins to see its outputs.

One criticism[1] that is often made of prototyping is that the requirements phase of a project using prototyping is too expensive, and keeping the requirements and design phases from running into each other is too hard. If those who are prototyping do not know what is needed for the prototype nor when they may begin building it, they may waste time and resources.

Why define reports and screens on paper? Why not go immediately to layouts or sample reports with a report generator? Often a person from Information Systems must start a user's thinking process by suggesting a layout. But, doing so is difficult for the Information Systems person if he has no idea of the contents, sequence, totaling, and uses of the reports. Even if the users will produce the reports with a report generator, beginning with the report generator now would be premature. You still need a prototype database with some sample data on it, before work with a report generator is meaningful.

Report generators and screen painters will be useful in building the prototype later, but what you need at this point is software for building data dictionaries and structuring databases.

CONTENTS OF THE PROTOTYPE DEFINITION

What will the prototype initially be and look like? It will be a working model of the system that will include the major program modules, the database, the essential screens, the reports, and the interfacing inputs and outputs used to communicate with other systems. It will be a skeletal version of the system and will not contain all the processing and validation rules that the system will finally have.

Table 4-1 lists the contents of the prototype definition. The definition has two parts:

1. What the system should do
2. How it should do it

The logical definition describes what it should do. The physical definition, which some call the physical design, describes how it should do it. Neither definition should be complete and exhaustive, since you should not try to make the initial prototype complete and exhaustive.

TABLE 4-1 Prototype Definition Contents

Logical Definition

- Reports
- Screens
- Information
- Functions
- Controls
- Interfaces

Physical Definition

- Database
- System Flow

The system whose prototype you are defining may be one of the following:

- New on-line system
- New batch system
- Package modification
- Enhancement to existing system.

The definitions of both a batch and an on-line system will be similar. The documentation of packages and existing systems should be adequate as a basis. The project team should only need to add definitions for those parts of the enhanced or modified system that are new or different. One exception is when the existing system or package does not have a well-defined database and the new system is a significant modification. Then, you should consider defining the entire database as part of the prototyping effort.

Logical Definition

The logical definition of what the system should do includes a description of the reports and screens the system should produce, the functions it should perform, and the information it should require.

Reports The description of each report should include its purpose, a list of the information elements the report should contain, the source of the information, how many pages it may contain, the sequence it should be produced in, how it should be totaled, how often the report should be printed, whether special forms should be used, how many copies of it should be produced, and who should receive them. Table 4-2 lists the contents of report descriptions.

TABLE 4-2 Report Description

- Purpose
- Contents
- Sequence
- Volume
- Source of Information
- Totals
- Frequency
- Distribution
- Special Forms

TABLE 4-3 Screen Description

- Purpose
- Contents
- Source of Information
- Volume
- Sequence
- Relationship to Other Screens
- Access

Screens Systems use screens in conversations for entering information and responding to inquiries. The screens include menus that list options, input screens, screens that are partly input and partly output, and screens that display the responses to inquiries. The description of screens should include their purpose, a list of the information elements to be displayed on each screen, the sources of the elements, the sequence in which screens will be displayed, the number of screens that will be used during a time period, the relationship of different screens in a conversation, and who should be allowed to view the screen. Table 4-3 lists the contents of screen descriptions.

Information The description of information tells what useable data the system develops and maintains as a database. It should include a description of the information elements and the records which contain them. You should use some sort of automated data dictionary system for maintaining this information. The system may be a purchased package or something that you develop, possibly using a microcomputer DBMS.

The information element definitions you develop for the prototype definition will not be complete and exhaustive, yet you should take great care in developing them. The success of the prototype and the system will depend on whether it is based on a good initial understanding of the information elements. Even though both the users and the Information Systems people may think they understand the information as they begin the definitions, they may learn that they do not. Persevering to produce good definitions will pay dividends for the rest of the system development process.

Table 4-4 lists the contents of a complete information element description and specifies which parts should be done for the prototype definition.

TABLE 4-4 Information Element Description

CONTENTS	NEEDED FOR PROTOPYTE DEFINITION
• Element Name	• Yes
• Element Abbreviation	• No
• Characteristics	• Yes
• Element Definiton	• Yes
• Subelements	• Yes
• Processing Rule	• Yes
• Alias Names	• No
• Validation	• No
• Headings	• No
• Responsible for Changes	• No

The element name is the best name by which the information element is known. Although all the alias names do not need to be found and recorded during this phase, if you know of other names that are used for the element, record them.

The element abbreviation is a short form of the element name that is often used in programs and database definitions.

The characteristics of an element include its size, what kind of characters (alphabetic/numeric/special) it contains, whether it contains any decimals, and whether, if numeric, it is always positive or negative.

The definition of the element is a word description of it. This is similar to a definition in an English language dictionary.

Subelement descriptions, including the positions within the element that they occupy, are important for defining an element. Many elements contain subelements. For example, a telephone number element may consist of an area code, a prefix, a station number, and an extension.

Some elements have a processing rule which tells how to develop or calculate them. A processing rule applies to an element, if the element is always developed or calculated the same way, using the same elements as sources, regardless of what databases or tables it appears on. An example of a processing rule for an amount net pay element would be:

Amount Net Pay = (Sum of all pay and additions)
− (Sum of all deductions and reductions)

The validation rules for an element tell what its allowable values are. They are not necessary for the prototype definition, but easily

TABLE 4-5 Record Definition Criteria

- Every record must have a unique key
- No repeating elements within a record
- No meaning should be given to position within a record
- Each element in a record is identified by the whole key
- No element depends on any other non-identifying element
- No needless multiplication of elements

stated ones, such as validate cost center identification against a cost center table, may be helpful.

The headings describe the standard headings to be used for an element on reports and screens. Spending time to develop them for the prototype definition is not advisable, so choose something quickly.

The identification of the organization responsible for changes to the element definition is also a part of the final element definition, but it is not needed for the prototype definition.

Table 4-5 lists the criteria for defining a record. Adhering to these criteria will help you develop a third normalized form of the record definition. Gane and Sarson[2] provide more information on how to develop this.

Table 4-6 lists the contents of a complete *record description* and specifies which parts should be done for the prototype definition. The record characteristics describe the record itself. The record contents describe the information elements in the record.

Every record must have a unique key. Define the record so that no two occurrences of the same type of record would have the same value for the key.

No elements should repeat within a record. No record definition should allow an element to occur more than once for each record. Analysts have a natural inclination to define records that contain twelve monthly summaries. They might be earnings for each month of the year. The first one would be for January, for example, and the third one for March. Doing so violates this rule and the following one that no meaning should be given to relative position within a record.

No meaning should be given to relative position within a record. The value an element may have should depend on its definition and its key, not on what happens to be next to it.

Each element in a record is identified by the whole key. A corollary to this rule is that no key should apply only to one or more elements.

TABLE 4-6 Record Description

CONTENTS	NEEDED FOR PROTOTYPE DEFINITION
• Record Characteristics	
Name	Yes
Estimated Number	No
Trigger for Preparation or Update	Yes
Allowable Response Time	No
Retention Period or Disposition Criteria	No
Responsible for Changes	No
• Record Contents (Information Elements)	
Sequence	Yes
Element Name	Yes
Source	Yes
Constant Value	Yes
Whether Standard Definition Applies	No
Whether To Be Edited	No

When the analyst in the example above realizes that he cannot have twelve earnings elements differentiated by position, he may want to distinguish them by a month code. The month code would further identify them, but would not apply to the other elements on the record, thus violating this rule as well as the one against repeating elements.

No element should depend on any other non-identifying element. If it does, the non-identifying element probably should be an identifying element. The analyst in this running example might have first viewed the month code as a non-identifying element, because in his mind it was clearly not part of the record's key. In one last attempt to keep the earnings summary elements on the record, the analyst may define them as twelve different elements with names like "Earnings Summary Month One" and "Earnings Summary Month Two." Doing that violates the rule that you should have no needless multiplication of elements. It also adds unnecessary elements differing from each other only trivially to the data dictionary.

Table 4-6 lists the contents of a record description. The description of the record characteristics includes the name of the record.

An estimate of the number of the records that will be used in or produced by the system will aid the final design of the system, but you do not need this estimate for the prototype definition.

The trigger for producing or updating a record is important information. If the record is an employee paycheck, the trigger may be a scheduled processing date. If the record is an employee profile, the trigger may be the receipt of an inquiry on an employee. Include the triggers for producing records in the prototype definition, if you know them.

The allowable response time for producing a record after the trigger occurs establishes service constraints for the system design, but is unnecessary for the prototype definition.

The retention period or disposition criteria tell when or for what reason records are removed from active status. This is not needed for the prototype definition.

The organization responsible for changes to the record description should be a part of the record description, but it is not necessary for the prototype definition.

The names of the elements on the record should also be a part of the prototype definition.

Records are organized by keys composed of information elements, so the sequence is a function of the elements on the record. Identify as keys all of the elements that would be logically required to sequence a record, regardless of whether they would physically be a part of the record when it is implemented in a DBMS. For example, in Figure 4-1 Employee Number would be shown as a sequencing element in the Earnings Record, even though it might not actually be used if the physical implementation were in a hierarchical DBMS. This information is necessary for the prototype definition.

Describe the source of each element on output or updated records. An element is either unconditionally posted unchanged from another record or it is placed there by a process, including conditional posting. You should identify the source of unconditional posting for the prototype definition, but you do not need to define in detail the process for producing it.

Sometimes certain elements always have the same value when they appear on a record. For example, if the element is a transaction code, it may always have a value of "CAY" on a certain record. If that is the case, that constant value should be defined as part of the prototype definition.

The standard definition, used to edit an element on inputs, may allow for a large number of possible values, but the element may have a more restricted set of values on the particular record you are describing. For example, a two-digit numeric element may normally have any value from "00" through "99." But, when it appears on the record you are describing, it will only have values between "00" and "10." If it has a more restricted use, then the standard definition does not apply for a particular record. You do not need to define that for the prototype definition.

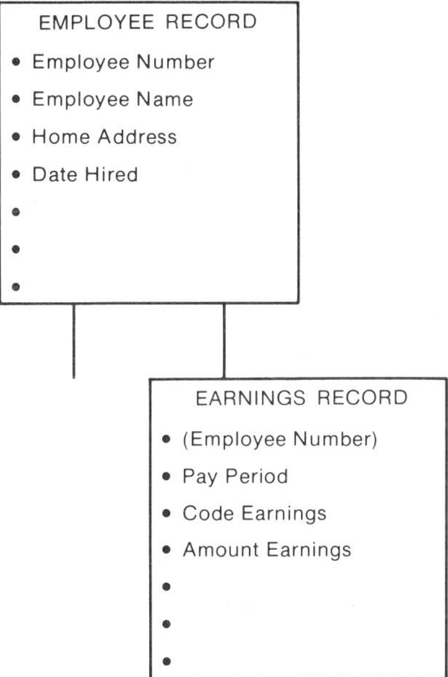

FIGURE 4-1 Partial payroll database structure

Whether or not each element on an input should be edited is an important part of the record description, but it is not necessary for the prototype definition.

Functions The description of the functions the system should perform should include a description of what the system should do, with references to the information in records that it should use and produce, and any outputs, such as reports and screens, it should produce.

A function description must at least be adequate as a caption for a function symbol on a data flow diagram. The description should sufficiently identify the symbol for those working on the prototype definition. The description must also be adequate as a title for a "black box" on a subsequent system flow diagram. The designation of the function's inputs, the records it will use, and the outputs it will produce must also be clear.

An example of a function description may be "Calculate Net Pay." Its inputs may include records containing recurring and one-time deductions and earnings. It may use payroll master records. Its outputs may include paychecks, check stubs, payroll registers, and updated payroll master records. Good notes on the present system's way of doing it may have been included with the data from the study of the present system. They may be useful as a detailed description of it. But, whether more detailed notes are available or not, a clear caption and specific identifications of information and outputs are enough for this part of the prototype definition.

Controls The definition of the controls should describe the controls on input, the program-to-program controls, and any controls for output. Even though the controls should not be implemented in the initial model of the prototype, "pegs" for them should be left in it.

Controls on input coming from people are usually batch controls. Controls on input coming from an interfacing system are usually totals that must be agreed with. Controls on output going to people are usually control totals. Controls on output going to interfacing systems are usually totals that the interfacing system must agree with. Program-to-program controls usually specify elements to be totaled in one program and compared in another and instructions on what to do if the totals do not agree from program to program.

Interfaces The description of the interfaces describes the inputs to the prototype from feeding systems and the machine-readable outputs it should produce for the use of other systems.

Physical Definition

The physical definition of the prototype is the design of the initial model of the prototype. It should include the database structure and contents and the system flow. The system flow should also include the design for prototyping whatever conversion will be required from the former system to the new one.

Database The definition of the database structure and contents depends on the record descriptions, the choice of records to be grouped together into a database, and whether you will eventually replace the database you use for the initial model of the prototype with a another implementation for the final version of the system.

You should not need to modify the record descriptions you prepared as part of the logical definition. Preparing the database definition from them may involve no more than deciding what DBMS to use for the initial model and whether grouping the records will require any additional elements. The frequency, volume, and sequence of reports and inquiries (screens) the system will produce from the database may also influence your final definition.

If you will group two or more records into the same database, each record should already have a unique key to distinguish it from any other occurrence of the same record. Nevertheless, you should determine whether you need to add key elements to distinguish one record type from another.

If you will use a different DBMS strategy for the initial model of the prototype, that is what you should define now. You will develop the final version as part of the prototyping process itself. For example, you may expect to use a hierarchical implementation for the final version of the database, yet you might initially define the database as overlays on a fixed-length, indexed file. The fixed-length, indexed file is what you need to define first. Note that you may need to add an additional element or elements to distinguish the different record types that will be on the same database.

System Flow The system flow defines the structure of the prototype. It should show the movement of information within the system from inputs and interfacing systems, through programs, the database, and program-to-program files, and to outputs and interfacing systems. It should depict the sequence of all program-to-program files.

The system flow should also describe any conversion that will be required from an existing system to the new system. The conversion process should be built as part of the prototype and prototyped along with the new system.

The system flow should identify the database records, the program-to-program files, the screens and reports, and any interfacing systems. It should show each program that will process input either from outside the system or from another program, each program that will produce output, and each program that will use or update the database. A program's description on the system flow may be the only description of it. Its definition should be that of a "black box" which moves data around. A caption for the box should identify the function. Insofar as possible, you should also identify the individual modules within each program and the functions they will perform. A programmer should be able to write the command language for the prototype from the system flow, so its identification of the parts of the system must be specific.

The description of the program-to-program files will include an identification of each file and a record description for each record it will contain. See Record Descriptions above and Table 4-6 to learn what should be in a record description.

ESTABLISHING THE PROJECT TEAM

The project team responsible for the prototype definition will be composed of users[3] and persons from Information Systems. The users who participate will do so because of their positions and what they know, not because of their skills in such things as requirements definition and system design.

The users should include the person who requested the system, the managers of the areas of the organization that will use the system, and the supervisors and other user specialists who understand the processing in that area of the organization. If they are good at visualizing new processing and expressing their needs, they will greatly aid the prototype definition process. If they are not good at those things, they may still help prepare an adequate definition. Remember that one deficiency prototyping is intended to remedy is the inability of many persons to visualize how something should work before it has been built.

The managers who establish the need for a report or a screen should say what it will be, what will be on it, and how it will be used. Similarly, the managers responsible for having work done should identify the functions the information system is to perform. The managers and supervisors who will be responsible for working with the system or producing data for it should understand what will be needed and why.

Table 4-7 lists the skills required for the Information Systems persons who will work on the prototype definition. The table does not imply that at least three persons from Information Systems must work on any prototype definition. One person may possess all of the listed skills. Or two persons between them may have all of them. Also, the table shows the management of the project as an Information Systems function, but in many organizations a user may manage development projects.

Even if the overall manager of the project will be a user, Information Systems management must choose the person whom they will view as the Information Systems project manager or project leader. This person should have the skills listed for management in Table 4-7. Whether he is also a business systems analyst, a systems analyst, a programmer, or a systems designer should depend on the other duties he might have on this project.

Until now different persons within Information Systems may have worked on the system request and the study of the present system. But, it is essential for the prototyping methodology that one or more persons stay with the project beginning with the prototype definition and continuing through the implementation of the resulting

TABLE 4-7 Required Information Systems Skills

MANAGEMENT

- Project Management
- Project Leadership
- Communication Facilitator

REQUIREMENTS DEFINITION

- Ability To Work with Others
- Interviewing Skills
- Definition of System Function
- Information and Database Definition
- Knowledge of System Area
- Knowledge of Commercial Systems
- Familiarity with Current Application Systems Trends and Techniques

PHYSICAL DESIGN

- Knowledge of Installed and Planned Hardware and Software
- Knowledge of Current Hardware and Software Trends and Techniques
- System and Communication Design

system. Thus, the most important consideration in picking persons from Information Systems is to select one or more who will provide continuity for the project through its life.

User management must also identify those managers and supervisors who will participate in the prototyping process from definition through implementation. If it is their responsibility, user management should also designate someone as their project leader or project manager, or even as the overall manager of the project. User management may also pick some managers and executives to serve with Information Systems management representatives on an executive review committee to review the progress of the prototyping project.

DEVELOPING THE LOGICAL DEFINITION

The first part of the prototype to be defined is that part called the logical definition. It is the definition of the prototype's functions, screens, reports, and information. Table 4-8 shows the tasks necessary to do the logical definition.

TABLE 4-8 Logical Definition Tasks

TASK	TO BE DONE BY	TO BE REVIEWED BY
Talkthrough system	Entire Group	Management
Define screens	User-Analyst Teams	Entire Group
Define reports	User-Analyst Teams	Entire Group
Describe functions	User-Analyst Teams	Entire Group
Define controls	User-Analyst Teams	Entire Group
Define interfaces	Analyst	Entire Group
Define information	Useer-Analyst Teams	Entire Group

System Talkthroughs

The logical definition should begin with a system talkthrough of the prototype. A system talkthrough is a series of discussions by all the involved users and Information Systems persons of what a system will do, use, and produce.

Table 4-9 shows the results you should obtain from a system talkthrough. The data flow diagram will briefly show the functions and the information flow. Additional notes on processing details that were discussed and the frequency of doing each function should accompany the data flow diagram.

Lists of outputs, including reports and screens, are another important result of system talkthroughs. If the discussion of reports encompasses their contents, sequence, and totaling, you should include notes on these things. But, the most important aspects of reports are their purpose, the frequency of producing them, and who should receive them. Likewise, if the discussion of screens identifies contents, at least partially, include notes on the contents as part of the results. Nevertheless, the most important information about screens is their purpose, frequency of processing, who should be allowed access to the information on them, and their relationship to other screens in on-line conversations.

System talkthroughs enable all the members of the group that are responsible for the new system to work as a team in defining the elementary requirements of the prototype. Everyone can hear the users listing what is to be done and Information Systems' suggestions of what to use, such as screens, reports, and the database. These system talkthroughs also establish group identity and help the team members learn to work with one another.

A system talkthrough should be two to three hours in length. More than one session will probably be necessary to define everything.

TABLE 4-9 Results of System Talkthrough

Logical Data Flow Diagram showing for each function
- Function Title
- Input Identification
- Output Identification
- Database Identification

Notes on each function, including
- Frequency of processing

List of reports and other outputs, including
- Purpose
- Frequency of processing
- Distribution
- Contents (partial)
- Sequence (partial)
- Totals (partial)

List of screens
- Purpose
- Frequency of processing
- Contents (partial)
- Relationship to Other Screens
- Access Allowed

List of interfaces with other systems

List of controls

No more than three weeks should elapse between the first and the last session. The group should meet as often as necessary to finish in that time period.

The group should meet in a comfortable conference room large enough to enable everyone to use materials they may bring and to allow the use of visual aids.

During a talkthrough everyone should be able to see what is being identified. Recording notes on a flip chart, a blank overhead projector foil, or a blackboard will allow this. Also, someone, including possibly a secretary or someone else who is not a member of the group, should prepare notes on the discussions in each talkthrough. The group should review the notes from each previous session at the beginning of the next one.

The project leader should play the role of the discussion facilitator. He should be responsible for guiding the discussions to the intended results. One approach which may help is concentrating individual sessions on specific topics.

The first topic for discussion might either be output or functions. The other topics might be input, controls, and interfaces.

Even the best project leaders have trouble focusing the discussions of a group on a constrained topic, such as output, especially when the members of the group are meeting together for the first time. The most the project leader may be able to hope for in the first one or two talkthroughs is a wide-ranging and somewhat free-wheeling discussion which hits on the principal functions and outputs of the system. The project leader should actually encourage such discussions. They may be among the most productive of any of the talkthroughs. Nevertheless, without being too forceful, the project leader should keep the discussions businesslike and ensure the participants are aware of the functions and outputs that they are identifying.

A good way to begin the first talkthrough is to ask the participants what they want the system to do or what they imagine the system doing. As the participants mention ideas the project leader should list them for all to see. He should guide them into listing as many things as possible.

After the group has spent its original enthusiasm in listing the main functions and outputs, the project leader should categorize what they have said. They have probably mentioned functions, reports, screens, and possibly even interfaces with other systems.

He should make sure the group completes its discussions of each. Naturally, to talk about functions without mentioning outputs or to talk about outputs without mentioning functions is almost impossible. Similarly, the discussions will disclose information on inputs, controls, and interfaces. The project leader must check that notes are taken when new information is mentioned and that all items are adequately discussed.

As the participants discuss the system's functions and its information flow, the project leader should sketch the results as rough data flow diagrams. Documentation assistants should make neater data flow diagrams for review at the next session.

As the discussions move to screens and reports the project leader will list the information that is developed. He should also show what information should be developed for each.

As source material, the talkthrough participants should draw on the service request and their knowledge and experience. Also, the material that was produced in the study of the present system should be useful as a source for discussions and as a checklist for making certain that all the functions of the present system are considered.

When the talkthroughs are finished, the project leader should review the results of the system talkthroughs with both user and Information systems management, either separately or together.

Describing Screens, Reports, and Functions

The entire project team is responsible for the completion of the screen, report, and function descriptions. But, after the talkthroughs are finished, the team may break into individual user-analyst teams to prepare the detailed descriptions. The limit to the number of teams would be the limit of the number of analysts or users or functional areas involved in the project.

Naturally the skills and experience of the individuals are also important factors in deciding what they should work on. Users who will use reports or who are familiar with the information that will be included on them should participate in the report descriptions. Analysts who have a feeling for screen design should participate in the screen descriptions.

Since the descriptions are ultimately the responsibility of the entire group, the entire team should meet at least once a week to review the descriptions that the various teams have developed.

As was mentioned earlier, screen-painting and report generating should not be done now. Team members who have used advanced tools for painting screens and generating reports may want to use them rather than bothering with paper. But, to do so will be a distraction at this point in the project. If they use them, they will want to test the reports and screens by setting up some test data, and the prototype building phase will have begun. This is the last chance the team will have to plan the system in a relatively calm atmosphere before beginning the more active parts of the prototyping methodology. Enjoy it.

Data Dictionary Entries

A small user-analyst team should develop the data dictionary entries. Although they should have the best knowledge of the information to be defined, they should also draw on the knowledge of other members of the group and even of persons outside the group as they work.

But, why dawdle over the data dictionary? You will know what should be in the reports and screens. Why not just throw something together for the database and start building the prototype?

Prototyping will be chaotic unless it is based on a good knowledge of the information. You should understand at least 80% of the content and structure of the data. If you do not have that before you build, you will have to make too many drastic, frustrating changes in the prototype. The effort may well fail.

TABLE 4-10 Data Dictionary System Features

- Provides all the contents listed for Information Element Descriptions in Table 4-4
- Provides for record and file descriptions
- Enables generation of coding from the data dictionary itself
- Enables finding all uses of elements and records
- Allows for versions of elements and records

So, take the time while you are still in the definition phase to develop a good understanding of the information the system will process. Developing this understanding and recording it is called making the data dictionary entries.

Should you actually use some kind of data dictionary system? If you can, yes. The data dictionary system may be a purchased package or a system you develop. Ideally it will contain the features listed in Table 4-10. A DBMS on a microcomputer is useful for developing your own data dictionary system, but including all the desirable features may be too difficult when you build your own.

A good data dictionary system should allow you to record a thorough description of each Information Element. This should include all that is listed in Table 4-10 for prototyping.

The data dictionary system should let you show the grouping of elements together into records and the grouping of records into files or databases.

You should be able to generate coding for at least the definition of the database, if not also for some of the processing, such as posting and validating, from the entries in the data dictionary.

The data dictionary should also enable you to find all the uses of elements and records in both programs and DBMSs.

To be useful in a production environment, a data dictionary should allow the recording of different versions of elements and records that were effective for certain periods of time.

If a data dictionary already exists, it should be reviewed to determine whether it contains entries for files that will be affected by the prototype. If you are modifying an existing system, it certainly should. The team working on the data dictionary should modify and augment any entries that exist. The entire project group should also review the data dictionary entries that the user-analyst team develops.

DEVELOPING THE PHYSICAL DEFINITION

The physical definition as shown in Table 4-11 includes the database structure and contents and the system flow for the prototype of the system and that of the conversion process. The physical definition is that of the initial model of the prototype; it is not the definition of the final version of the system. The initial model will be the vehicle for establishing the database, painting screens, and generating reports as the project team prototypes.

If the initial model will be developed with different tools than the later version of the prototype, design it for the tools that you will use first. The use of the other tools will be adopted as the prototyping progresses.

One or more design analysts and database administrators, who may not be members of the project team, should work with the Information Systems project leader to produce the physical definition. They should do it after the completion of the logical definition and should base it on the logical definition.

After they have completed the design they should do a walkthrough of it with the other team members and anyone else designated by the organization to participate in the walkthrough of a new physical design. The purpose of the walkthrough is to verify that the design provides a good framework for the initial model of the prototype and that it meets the organization's design standards.

TABLE 4-11 Physical Definition

- Database Structure and Contents
- System Flow
  - Prototype
  - Conversion

DOCUMENTING THE DEFINITION

The project leader should make sure that copies of the written results of the logical and physical definition of the prototype will be maintained as a part of the project records. The database will change. The reports and their distribution will not be stable. The screen descriptions will not be the last word. Yet the completion of the written definitions is a critical milestone in the project and maintaining a copy of them is important to the history of the project. See System Documentation in Chapter 7.

PLANNING AND PROGRESS MONITORING

The definition of the prototype establishes the basic requirements of the new system and the framework of the prototype. Establishing a plan for the definition and monitoring adherence to the plan will ensure a good definition is produced with a reasonable expenditure of resources. Table 4-13 lists the tasks that comprise the definition. You will find a list of elements considered essential to the success of the phase in Table 4-12.

Securing the cooperation and enthusiasm of team members is the responsibility of the project leader with the help and support of both user and Information Systems management. The team members should identify themselves with one another and with the project and should accept responsibility individually and as a group for the new system's success. To accomplish this the project leader should make certain that all the team members understand what they must define in this phase and the steps they will take to do it. He should also discuss with them the different tasks they may work on to complete the definition.

The team will identify the system features in the talkthroughs before doing the rest of the definition. The framework they produce will be the basis for all that follows in this phase as well as the remainder of the project. Again the project leader must be responsible for creating an atmosphere in which the team members can work productively.

If the team does not understand the database and define it well, the prototyping will certainly flounder, if it does not fail. The results of the screen, report, and function definitions should prompt the information definition, even though they can proceed at the same time. The team members who are responsible for the definition of the informa-

TABLE 4-12 Essential Elements for Definition Success

- Secure cooperation and enthusiasm of team members
- Identify the system features in the talkthroughs
- Understand the database
- Follow a firm schedule
- Monitor quality and quantity of work

TABLE 4-13 Definition Tasks To Be Scheduled

Logical Definition Tasks

- Talkthrough system
- Define screens
- Define reports
- Describe functions
- Define information
- Define controls
- Define interfaces

Physical Definition Tasks

- Design database structure and contents
- Design system flow
- Walkthrough physical definition

tion should be persistent, thorough, and able to judge the completeness of their results.

As mentioned in Chapter 3, estimate the work to be done, plan it, schedule it, publish the schedule, and adhere to it. You should not need to allow for iteration of the tasks in this phase. In fact, the team should complete the logical definition before the physical definition begins. But, after the talkthroughs are finished, you may define the screens, reports, functions, and information simultaneously, if enough persons are available. Since the number of reports and screens and the complexity of the information may not be known until after the talkthroughs, you may want to prepare a preliminary plan and schedule for this phase before the talkthroughs begin, then modify them after they are completed.

The project leader and both user and Information Systems management should monitor the quality, quantity, and timeliness of the work that the team produces.

SECURING EQUIPMENT FOR PROTOTYPING

If a system extends on-line processing for the first time to some part of the organization, new or additional equipment may be needed. Users will be exercising the prototype after it is built. They should not have to leave their area to work with the prototype. The ideal is to place terminals for prototyping an on-line system in the user area. The project team should ensure that equipment is installed[4] in the user area by the time the exercising of the prototype begins.

If sufficient equipment of the same type that will be used in the real system, or of a type close enough to it, will not be available when exercising the prototype begins, you should consider buying or possibly even temporarily renting equipment.

Also, if the new system will include communications with outlying parts of the organization, such as branch offices, equipment and communications should be provided in at least one nearby location for prototyping.

PLANNING FOR BUILDING THROUGH IMPLEMENTATION

The next three phases of the project include:

1. Building the prototype
2. Exercising the prototype
3. Implementing the system

Since you have defined the prototype, you should have a good basis for estimating what will be neededd to build it, exercise it, and implement the system that it becomes. If this is the first time you have prototyped, you may want to read ahead before completing your estimates and developing your plan. But, develop your first detailed plan for the next major phases of the project now, before continuing with the project. A schedule and cost estimates should accompany the plan.

Milestones

As you develop your plan, remember the importance of milestones to it. The project must have milestones—points in time when all agree that the requirements have been defined, the system has been designed, the acceptance testing haas been completed.

The project team must produce documents that describe the requirements and the acceptance test plan. The team may prepare these documents while the phase that they are to document is underway. What happens during the phase must conform to the document, or the phase cannot be considered to have been successfully completed. Neverthess, the beginning of the phase does not have to be delayed until the document is ready.

As an example, the testing must conform to the test plan. But there is no need to delay beginning the testing, while waiting for the test plan. All should agree, though, that the testing cannot be considered completed until the test plan has been completed. Only then can it be known whether the testing conforms to the plan.

How can the testing begin before the test plan has been developed? Testing is not an after-the-fact proof that something has been done; it's exercising the prototype. It's determining what the system should do, modifying it, putting test data through it to ensure it has been done, then modifying it again and testing it again. The test plan then is a description of this exercising.

SUMMARY

1. The prototype definition will be the basis for building the prototype.
2. Contents of the prototype definition:
   - Logical Definition
     Reports
     Screens
     Information
     Functions
     Controls
     Interfaces
   - Physical Definition
     Database
     System Flow
3. The logical definition describes what a system should do.
4. Record definition criteria:
   - Every record must have a unique key
   - No repeating elements within a record
   - No meaning should be given to position within a record
   - Each element in a record is identified by the whole key
   - No element depends on any other non-identifying element
   - No needless multiplication of elements
5. The physical definition of the prototype is the physical design of the initial model of the prototype.
6. The project team responsible for the definition will be composed of users and persons from Information Systems.

7. Approach for defining the prototype:
   - Talkthrough system
   - Define screens
   - Define reports
   - Describe functions
   - Define information
   - Define controls
   - Design database structure and contents
   - Design system flow
   - Walkthrough physical definition
8. A system talkthrough is a series of discussions by all the involved users and Information Systems persons of what a system will do, use, and produce.
9. Schedule the prototype building through implementation phases, paying particular attention to definite milestones.
10. Install equipment in the user area for prototyping.
11. Elements essential to the success of the definition:
    - Secure cooperation and enthusiasm of team members
    - Identify the system features in the talkthroughs
    - Understand the database
    - Follow a firm schedule
    - Monitor quality and quantity of work

FOOTNOTES

1. Mason, R.E.A. & Carey, T.T., "Prototype Interactive Information Systems." *Communications of the ACM*, 1983, *26* (5), 347-354.
2. Gane, Chris & Sarson, Trish, *Structured Systems Analysis: Tools and Techniques*. Englewood Cliffs, NJ: Prentice-Hall, Inc., 1979.
3. Chapin, Ned, "Prototyping—Quick, Not Dirty ..." *Data Management*, 1983, *21* (10), 46-48.
4. Booth, G.M., *The Design of Complex Information Systems*. New York, NY: McGraw-Hill, Inc., 1983.

Chapter Five

BUILDING THE PROTOTYPE

This chapter describes building the prototype, including the database, program modules, screens, and reports. It also discusses expanding the team, choosing tools, and building the prototype on a microcomputer.

In the phase described in the last chapter, you define the blueprint for the prototype by:

- Doing system talkthroughs
- Listing system functions, screen and report contents
- Defining the information using a data dictionary
- Developing the physical definition
- Ordering equipment needed for prototyping

In the prototype building phase you are actually going to put all the pieces together and make them work. The initial model of the prototype will be a software mock-up. It will move data between program modules and to and from communicating systems, process input from screens, and produce reports and screens. Table 5-1 shows what you must build.

TABLE 5-1 Building the Prototype

- Build the database
  Define it
  Enter test data into it
- Build the initial model
  Program modules
  Screens
  Reports
  Program-to-program files

TABLE 5-2 Criteria for Initial Model

- Bring it up fast
- Build all the components
- Move data with it

TABLE 5-3 Completeness of Initial Model

• Database	• Complete
• Test data	• Partial
• Program modules	• Partial
• Program-to-program files	• Complete
• Screens	• Partial
• Reports	• Partial

This chapter will also consider the differences in building the prototype depending on whether the system it is for is batch, on-line, based on a package, or an enhancement or modification to an existing system.

Table 5-2 lists the criteria for building the prototype. Speed is the most important consideration in building a prototype. Build something quickly so you may begin to exercise it.

Build all the components of the initial model. Construct a database from its definition as completely as possible. Also, make all program-to-program files as complete as possible. The other components of the prototype must all be present. You must program something for each program module, even if it is only an entrance and an exit. You must produce something for each screen or report, even if it is only a heading or one or two elements. You must have some data on the database. But, you do not need to complete all parts of the screens, reports, and program modules. Table 5-3 shows which parts of the initial model must be complete and which may be only partly done.

The prototype must work, or you cannot exercise it. Build the initial model so that it moves data from initial input through final output.

Expanding the Project Team

Chapter 4 discusses establishing the project team to define the prototype. Now you must expand the team, first to build the prototype and then to exercise it. Ideally the same users will stay with the project, al-

TABLE 5-4 Prototype Building Tasks

COMPONENT	TO BE DONE BY	
	WITH 4GL	WITHOUT 4GL
Database	User-PA Teams	Database Admin.
Test data	User-PA Teams	User-PA Teams
Program modules	User-PA Teams or PAs	PAs PAs
Screens	User-PA Teams	PAs
Reports	User-PA Teams	PAs
Program-to-program files	PAs	PAs

though those who are managers may not spend much time building the prototype or making the modifications to it that exercising it requires. Different Information Systems persons may join the project now. And, some of the Information Systems persons skilled in preliminary requirements definition may leave the project.

Persons with different skills should work on different parts of the prototype building. Table 5-4 lists who should build the different components. If a fourth generation language (4GL) is used, teams composed of users and programmer-analysts (PA) can build many of the prototype's components. If a fourth generation language is not used, programmer-analysts will have to do more of the prototype building. They will need to use problem solution languages and utilities which users cannot readily learn.

Table 5-5 lists the skills the team members need to build the prototype. Either users or Information Systems persons may possess some skills, depending on whether fourth generation languages are used. For example, if a relational DBMS is used, users may readily learn how to define database records. If a hierarchical database is used, not only may the users on the team not be able to define the database, but the programmer-analyst may also need the assistance of the database administrators. Similarly, the utilities for loading a relational database are often easy to use, whereas one to load a network database may require specialized Information Systems skills.

Team members should be skilled in the language or languages to be used for writing the program modules and the command language procedures that will comprise the prototype. Some team members should be skilled in structured programming and modularization techniques to ensure the modules and their interrelationships are well developed. Team members will also need knowledge of the screen painters or mappers and the report writers that will be used.

TABLE 5-5 Skills for Building Prototype

- DBMS Definition Language
- Database Loading Utilities
- Ability to Program in Language Used for Modules
- Structured Programming and Program Modularization Techniques
- Screen Painter/Mapper
- Report Writer
- Knowledge of System Requirements and Data Characteristics
- Procedures for Control of Different Versions

Knowledge of the system requirements and data characteristics is also essential for successfully building the prototype. Part of this knowledge should be provided by the users and the Information Systems persons who have stayed with the project to provide continuity. Another part of it should be provided by team members who are able to use and understand the prototype definition and data dictionary entries made in the previous phase.

Another function, which will be even more important when you begin to exercise the prototype, is maintaining control of the different versions of its components. The title for this function is exercising coordinator (see Chapter 6). Someone, possibly a documentation specialist or a programmer, should have this as their primary responsibility.

CHOOSING THE TOOLS

As Chapter 1 stated, you can prototype without using any of the tools commonly associated with it. But, using tools does make prototyping easier. Table 5-6 lists those tools that help you to prototype.

TABLE 5-6 Tools for Prototyping

- Data dictionary
- Interactive testing
- Version control
- Fourth generation language
- Relational DBMS

Data dictionary You should have already chosen the data dictionary to use when you defined the prototype. It should serve you as you build the prototype. For example, when you use the name of an element in painting a screen or specifying a report, the data dictionary ideally will provide you the characteristics of the element for incorporation in the screen map or report program. If the data dictionary will not do that automatically, you should at least be able to use it to get the definition and characteristics of each element that you use. If changes arise as you are building the prototype, the data dictionary should allow you to update it easily.

Interactive testing An interactive testing system lets you use a terminal to change programs easily, to submit frequent tests, and to review the results of test sessions quickly without having to wait for batch output. It should also allow you to switch readily between different versions of programs and databases. You will be testing extensively as you exercise the prototype, but you will also be testing as you build it to ensure that it works.

Version control A library system, which may be a part of the operating system, is important for exercising the prototype because it allows the storage and retrieval of different versions of the prototype's components. Even though you will not use it until exercising begins, you should prepare for it when you build the prototype. The components are all the parts of the prototype: program modules, screen maps, record definitions, report programs, collections of test data, and command language procedures.

You may have reasons for having more than one version of the prototype at any one time. If you do, what is said here for one version applies to each that you may be using.

You need to know two kinds of version information about the prototype:

1. The version of each component that was part of it any time it was operating
2. A description of each component.

Fourth generation language The features of a fourth generation language (4GL) that are particularly helpful are its report writer, screen mapper, and query language. Fourth generation languages often contain their own DBMS. You may actually obtain these features separately without obtaining a fourth generation language.

You may want to use a fourth generation language for parts of the prototype or for all of it. You may want to use it temporarily or also in

TABLE 5-7 Fourth Generation Language Features

ADVANTAGES	DISADVANTAGES
• Quick Implementation	• Stylized Formats
• Easy To Use	• Requires Own Environment
• Easy to Change	• Peculiar Files
• User-Friendly	• Complex Logic Is Difficult
	• Inefficient

the final version of the system. In deciding whether and where to use a fourth generation language, you should consider the advantages and disadvantages listed in Table 5-7.

Fourth generation languages enable you to bring up the more straightforward parts of the prototype quickly. Establishing databases, mapping screens, writing reports, and setting up inquiries are easy using them. Making a change this afternoon to what you did this morning is easy. With guidance from Information Systems programmer-analysts users normally learn their basic features readily and can work extensively with them.

Fourth generation languages often have stylized formats for screens, reports, and inquiries. You may not want to continue to use these formats if you replace the fourth generation language. Since users and Information Systems persons may have become familiar with the formats of the fourth generation language, some additional time may be needed for familiarization and training with the ultimate formats.

Fourth generation languages often require their own environment. If a mixture of languages is allowed at all, the fourth generation language may have to be in control. These languages often use only their own peculiar files. These factors may require that major portions of the prototype be controlled by fourth generation languages.

Programming simple processes in fourth generation languages is usually easy, but programming complicated logic may be much more difficult than with problem solution languages like COBOL.

Programs developed with fourth generation languages are often operationally inefficient. This should not affect the use of fourth generation languages for the early versions of the prototype, but it should influence the decision of whether to keep parts of the final version in the fourth generation language. You should balance estimates of processing cost differences during a time period, such as one or two years, against any increased development and maintenance costs using a problem solution language.

You must also balance the ability the fourth generation language gives you to get the prototype up fast against the later need to reprogram, and possibly to redo the physical design for the use of the ultimate language or the ultimate DBMS.

Relational DBMS Using a relational DBMS for prototyping lets you modify the database much more quickly than using a hierarchical or network DBMS. You must balance the ease of using it with the work that will be required to replace it with the ultimate DBMS.

PLANNING TO BUILD AND MONITORING PROGRESS

Table 5-8 lists the essential elements for success in building the prototype. The plan and schedule that you develop should meet the criteria for the initial model in Table 5-2 and should include a firm schedule with specific assignments for the various team members. Earlier parts of this chapter discuss what must be done to build the prototype. Table 5-4 lists the prototype building tasks.

Table 5-3 differentiates between those components of the prototype that should be completely built and those that only need to be partially built. You do not need to develop a set of test data that will exhaustively test the prototype's features, but you should develop some data for all the different information elements and records on the database. Whether you should completely build a screen should depend on whether the screen is significant for modifying the database or controlling the processing of the system. Similarly, whether you should completely write a report should depend on the importance of the report for reviewing the contents and status of the database.

One criterion for building the prototype is to bring it up fast. If the team is large enough, on a typical day, early in the prototype building phase, some team members will be specifying the database, user members will be developing test data to load into the database, teams of users and programmer analysts will be painting screens and writing

TABLE 5-8 Essential Elements for Prototype Building Success

- Assign specific responsibilities
- Publish and follow a firm schedule
- Keep team members and management informed
- Monitor quality and quantity of work
- Follow the criteria for the initial model (Table 5-2)

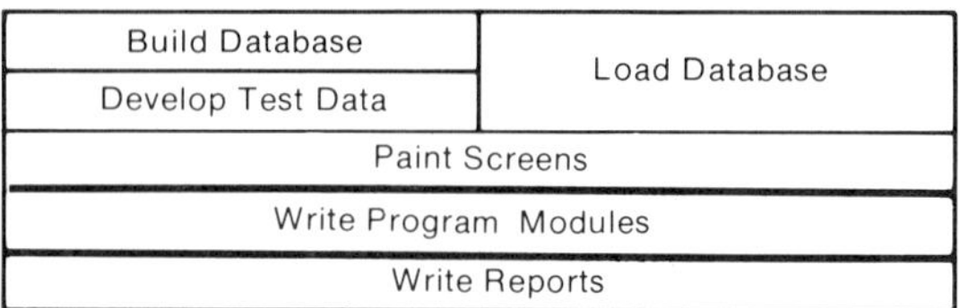

FIGURE 5-1 Simultaneity of prototype building tasks

reports, and others will be developing program modules. Figure 5-1 depicts this overlapping of tasks.

Drawing a line between building the prototype and exercising it is difficult, because you should verify that the modules function correctly before you are finished. This requires testing them. If they do not perform properly, you should correct them. This implies iteration which is the very life's breath of exercising the prototype. Nevertheless, to be methodological about prototyping, you must schedule distinct beginnings and endings to the various prototyping phases, including building the prototype. Likewise, you must assign persons to individual tasks with the intention of completing these tasks with the manpower allocated by the dates indicated.

The project team members should meet frequently to discuss progress and problems. The project leader should also meet often with user and Information Systems management to inform them about the project.

One final key milestone should be a demonstration of the initial model to user and Information Systems management at the completion of this phase.

THE INITIAL MODEL AND FINAL SYSTEM

The initial model of a prototype should *feel* as much as possible like the final version of the system. This is why you should expedite securing any special equipment that may be needed, as mentioned in Chapter 4. But, what does feeling like the final version of the system mean? If a specific brand of CRT terminal will be used for input, should you only use such a terminal for prototyping? If a report will be produced on special stock paper, should you wait for that paper before running the report? The answer to such questions is "Yes;" it is also "No." Before you finish prototyping, you should use the terminal you will use in the production system and you should run the report on the special form for it. Nevertheless, in the early stages of prototyping you may use other terminals and plain stock paper.

TABLE 5-9 Guidelines for Building Initial Model

DON'T CHANGE:

- Media of external inputs or outputs
- Sequence of inputs, outputs, or program-to-program files
- Structure of the system or interrelationships between programs
- Database sequence, contents, or structure

YOU MAY CHANGE:

- Media of database or program-to-program files
- DBMS
- Programming language and command language
- Specific input or output devices or forms
- Computer

You should not simulate a report with a screen, nor should you simulate a screen with paper. But, you may simulate one terminal with another or one type of paper with another. Table 5-9 specifies what you should or should not change in the initial model of the prototype compared to the final version of the system.

Since the prototype eventually becomes the system, all differences between the initial model and the final version of the system must gradually be changed. You should weigh the work of making the changes with the advantages creating temporary differences will give. Table 5-10 lists the possible advantages you may receive from making such differences.

A difference in the initial model may be much easier to use than what will be used in the ultimate system. A relational DBMS may be much less complicated than a hierarchical one. Running tests with a program-to-program file on disk rather than the tape that will later be used for it should improve the turnaround time for tests.

TABLE 5-10 Possible Advantages of Temporary Differences

- Easy to use
- Start up quickly
- Easy to modify
- Available
- Accessible

You may be able to start up the initial model more quickly by making some differences between it and the final system. An obvious example is not waiting for special forms before printing reports. Another example is that bringing up program modules with a fourth generation language is much quicker than with a problem solution language.

Some differences in the initial model may make it easy to modify compared to the ultimate system. Changing a relational database should be much quicker than changing a network database.

Some equipment may be more readily available than the exact equipment that will be used in the final version of the system. This may be true of storage media or input and output devices, or even computers.

Also, some equipment may be more accessible than what will eventually be used. You may be able to use microcomputers, but not able to schedule sufficient time on a mainframe like what the final version of the system will use.

Building the Prototype on a Micro First

Even though a new system will ultimately operate on a mainframe or minicomputer, you may have cogent reasons for building it first on a microcomputer.

As an example the new system may be for casualty insurance claims administration. It will contain provisions for recording new claims, evaluating them, assigning them to claims administrators, processing them, informing reinsurers of them, paying them, closing them, and reporting and answering inquiries on them. Bringing up the system will require adding records and information elements to an existing insurance database on a hierarchical DBMS containing information on insurance policies in force, premium payments, and reinsurance. The system will be on-line on a large mainframe using dumb terminals for updating and inquiry.

The system will consist primarily of COBOL programs with some infrequently used screens and reports in a fourth generation language.

Some problems you might encounter include:

1. The claims department has no terminals that may be used for prototyping and delivery of additional terminals will require six months.
2. Data administration does not want to set up a new test database, when the team is only 80% sure of its contents and structure.
3. Data administration does not have the resources to make frequent changes to the database during the early prototyping.

You might have available:

1. Several microcomputers.
2. A COBOL processor with provisions for indexed sequential files.
3. A microcomputer version of the fourth generation language you will use in the final version of the system.
4. A relational DBMS peculiar to microcomputers.

You could install microcomputers in the claims department. They would serve as terminals for the early stages of prototyping. You might write the program modules for the initial model in COBOL and the fourth generation language. You would change the media of the database and program-to-program files to floppy diskette.

To simulate a hierarchical database you might use overlaying definitions of different record types on an indexed sequential file.

Naturally, you would be temporarily changing computers, command language, and printers.

Using microcomputers this way is within the guidelines for building the initial model in Table 5-9. In fact, you would not have changed programming language.

A different alternative would be to build the initial model using the DBMS peculiar to the microcomputer, instead of using COBOL. This might enable you to shake out at least some of the changes to the database more quickly. But, since this might be quite different from the final version of the system, you might not want to build the complete initial model. Furthermore, you might not want to exercise the prototype much before rewriting the programs.

BUILDING THE DATABASE

The following is the definition of database used in this book:

> *database: a file used by more than one system and accessible in more than one way*

The definition of file follows:

> *file: a collection of one or more record types*

According to these definitions, an indexed sequential file with alternate indices would be a database if it is used by more than one system. It is a collection of records, accessible in more than one way, and used by more than one system. Its DBMS may be no more elaborate than the implementation of its access method.

Building the database is specifying it to the DBMS. The definition of it in the data dictionary is the source for its specification. Ideally, the actual specification statements may be generated from the data dictionary entries.

The project team is responsible for the database specification, whether it is generated from the data dictionary, by persons working on the project, or by the database administrators. The database to be built may be either hierarchical, network, relational, or indexed. If it is hierarchical or network, designers, possibly from the organization's database administration area, should specify it. If it is relational, users and programmer-analysts may specify it. If it is indexed, programmer-analysts will need to apply their skills to specifying it.

If the prototype is for a new on-line or batch system, the database may be entirely new. If the project is based on a software package, it has probably already been specified. If the project is a modification to an existing system, the database may be a change to an existing database.

You should begin building the database as soon as possible and try to complete it as quickly as possible. You cannot load test data into it until you have finished it. And, without the database you cannot test program modules, screens, and reports that use it.

Especially if the DBMS used for the initial model will not be the one used in the final version of the system, build data access modules that insulate the programs from the actual DBMS. A using program will call a module to bring in or put out each record. Then the later migration to another DBMS should only require changing the access modules.

Entering Test Data

The early entry into the database of some selected test data will help you test the processing of the initial model and its generation of reports and output screens. Selected test data is small in quantity, but intended to contain a rich variety of combinations useful for testing the features of a system. Not only is an initial set of test data on the database a part of the initial model, but it is also needed to test that the initial model was properly built and does work.

The users on the project team should lead the development of the selected test data with the assistance of programmer-analysts. If easy-to-use, on-line utilities are used for loading the database, users should be able to do it. If specialized batch utilities are used, programmer-analysts on the team should probably load it.

Chapter 6 will discuss developing test data in more detail. But, you should save and label any transactions you use to test the initial model in this phase. They may form a part of one or more test data sets to be used to exercise the system.

BUILDING THE INITIAL MODEL (SKELETAL SYSTEM)

In addition to the database, the initial model of the prototype consists of program modules, program-to-program files, screens, and reports.

Program Stubs Moving Data

The prototype's ability to move data is one of its essential features. Since the prototype should be as realistic as possible, you should want the modules of the system to appear to be complete when viewed from the outside. As you exercise the prototype, you will concentrate on different modules at different times, changing them. To test them you will want the modules that communicate with them to do so in a realistic way. Consequently, whether an individual module is complete is not important; what is important is that the module appears to be complete—that it moves data. In this way, one may view a prototype as comprised of program stubs moving data.

Each module should appear to read all its input files and to produce all its output files. It may initially dummy out its output files with hard coding, until logic can be placed in the module to produce output from input and database data.

As an example, a payroll system module is supposed to produce the file from which the system will print the paychecks, pay stubs, and registers for everyone on a payroll. It calls a module to read a calendar specifying what payrolls are to be run on what days, calls other modules to find the database records for those employees to be paid, calls a module to calculate the net pay and produce the file for the checks, stubs, and registers. The initial model will read the calendar first. Then it will find the database records containing basic information on employees such as social security number, name, and home address. It will place this data on the output file along with dummy information representing the results of the payroll calculation. The team will add the full logic and calculations later in the exercising phase.

If the prototype is for a new on-line or batch system all the modules will be new. If it is for a system based on a software package, the only new modules may be those for a pre-processor or a post-processor to the package itself; the additional modules will join a test copy made

TABLE 5-11 Sources for Building Modules

- System flow
- Functions (Logical definition)

of the package for prototyping. If the prototype is of a modification to a system, many of the programs may not be changed. Test copies of the programs that will not change along with copies of any modified or new modules will comprise the prototype.

Table 5-11 lists the sources for building the modules. The system flow depicts all of the modules and their functions, shows the data they will use and produce, and refers to the program-to-program files. The logical definition contains a further description of those functions, referenced on the system flow, that seemed to need additional explanation.

Team members should write the program modules. Wherever a fourth generation language is used, user members, with the guidance and assistance of Information Systems programmer-analysts, may be able to write them. If problem solution languages are used, the programmer analysts will have to write them.

After they are written, test whether the modules can read their input and whether the output they produce can be read by the modules that receive it. If the module produces reports or screens (see below), determine whether they conform to the prototype definition. When you have successfully tested a module, leave it and go on to other tasks. Remember you are only building the initial model, not exercising it. Exercising the prototype will begin when the initial model is ready.

Painting Screens

Screens provide users the ability to communicate with information systems. With screens users can inquire about information, enter or update information, or receive information.

Painting a screen is the process of laying out the format of a screen, specifying the information that will be on it, identifying the database that will be the source or recipient of its contents, establishing the order of the conversation that uses the screen, and providing access controls. Table 5-12 lists the different facets of screen painting.

Establishing the order of the conversation includes specifying the sequence in which information will be displayed on a screen and the conditions for displaying or using other screens. Access controls are used to restrict who can enter or receive information with a screen.

TABLE 5-12 Facets of Screen Painting

- Format
- Contents
- Source/disposition
- Sequence (of presentation)
- Relationship to other screens
- Access

The screen description portion of the logical definition is the source for all the different parts of screen painting except the format and the information element descriptions. It also includes the determination whether any particular screen should be completely or only partially prepared for the initial model. The team members will lay out the format as they paint the screens. The data dictionary is the source for the information element descriptions. Ideally, extracting information from the data dictionary will be automatic.

Whether screen painting is easy depends on the tools to be used. Most fourth generation languages contain a screen mapper which simplifies the process. A person using a screen mapper lays out the format of the screen, differentiating between constant and variable data. He then identifies each of the variable information elements and the database record that contains them. He may also specify the sequence of the information, what screen the system should display next, and who may access the system using the screen.

If a fourth generation language is not used, the team members will paint the screens directly with a communications control language or indirectly with a map editor. A communications control language requires that the screen specifications be made in a problem solution language, such as a set of COBOL functions. A map editor allows the user to paint the screens somewhat as he would with a fourth generation language. Programs supporting the editor will then produce the actual communications control language statements that are needed.

A screen mapper is normally easier to use than a map editor, and a map editor is easier to use than a plain communications control language. The critical difference is in who can do the screen painting. If a fourth generation language is available, users on the team can do much of the screen painting with guidance from Information Systems members of the team. If a map editor is available, users may still be able to do much of the screen painting. If only a communications control language is available, then programmer-analysts must write the actual language statements; although users may assist, especially contributing to the screen layouts.

If, when planning to build the prototype, the decision was made not to build a screen completely, then the screen you build only needs an identification or a heading.

Example of a new on-line system As an example, the prototype may be for a new on-line automobile insurance claims system. A screen to be painted is the one on which a clerk records a claim when the company first receives notice of it. The clerk will record the claim while taking a phone call in which the claim is being reported or from a written notice made either by the claimant or an adjuster.

The decision made in the prototype definition phase was that this screen would be complete. Table 5-13 lists as input the data to be entered on the screen. Also, the system will produce a verification of adequate policy coverage as output.

A fourth generation language is available for screen painting. The data dictionary to be used can be accessed on-line, but the element definitions from it cannot be incorporated automatically into the screen painting process.

A claims department supervisor will have been trained to use the fourth generation language, but will not have much experience using it. The supervisor sits at the terminal, and a programmer-analyst sits with the supervisor.

The supervisor first does the format. The supervisor is particularly concerned that the layout of the screen fits the way the clerk should work and also fits any documents the clerk uses as sources. Then the supervisor and the programmer-analyst review the format and identify the various information elements and their sources. They use the data dictionary for the exact names and characteristics of the various elements. The programmer-analyst specifies how the system will call up the screen and what the access codes will be.

TABLE 5-13 Information for Logging Insurance Claims

• Claim number	• Input
• Claimant name and address	• Input
• Policy number	• Input
• Adjuster identification	• Input
• Description of the claim	• Input
• Date received	• Input
• Estimated claim amount	• Input
• Verification of coverage	• Output

Example of a package modification Another example is a prototype of a modification to a financial system package for general ledger, budget, and cost accounting. The modifications include using the package's input parameterization features to customize its processing of input from screens and various subsidiary systems, such as inventory control, billing, and accounts payable. The team will also use a fourth generation language to produce output screens and a stand-alone report writer to prepare reports.

An important input screen is the one used to enter manually prepared journal entries. The decision made in the prototype definition was to paint it completely. Its description is a source for the screen painting process. The package dictates the order of the conversation of which the screen is a part, how access control will be handled, and what information elements must appear on the screen. It allows you to decide on most major headings, the placement of elements on the screen, and their exact size and characteristics. It also allows you to add several additional characters of data which the package will maintain as part of the journal entry.

The screen painter is a map editor. A person uses it on-line to lay out the format of a screen and specify the size and characteristics of the information elements. The package then edits the screen map. If it accepts it, you may run it as a test.

The package has its own limited data dictionary which maintains changes to package element characteristics. The organization's data dictionary is the source for entries to the package's dictionary. The organization's data dictionary can be accessed on-line, but the element definitions from it cannot be incorporated automatically into the package's dictionary.

A supervisor from corporate accounting, who has been trained in using the map editor, will paint the journal entry screen with assistance from a programmer-analyst. The supervisor will first lay out the screen. Then, with the help of the programmer-analyst and drawing information from the organization's data dictionary, the supervisor will specify the information elements and their characteristics. The package has pre-determined their sources and dispositions.

Example of a system enhancement A final example is a prototype of an enhancement to an airline customer profile system to change it to an on-line reservations control system. The existing system contains many reports and inquiries on passenger trips and allows on-line input. The enhanced system will retain the reports and inquiries, buts its focus will change to supporting reservations clerks as they make flight reservations. The enhancement will operate on a microcomputer. The team members will program it in the same relational DBMS language used for the existing system.

The existing system contains a screen for entering information on a reservation. The screen in the enhanced system will be similar, but will have less elements and a modified layout.

A limited map editor is available. It will partially produce programs for displaying screens. Programmers have to expand the programs it generates.

Team members use the DBMS to implement a simplified data dictionary. It has no automatic link to the screen mapping process.

A programmer-analyst and the reservations department manager, who is a member of the project team, use the map editor to lay out the modified screen. The manager sketches what he wants on a screen, then the programmer-analyst adds element designations. They look up element characteristics in the data dictionary to add them to the screen map. The programmer-analyst writes a simple program to display the screen and to allow data to be entered on it. The program at first will not update the database.

Programs that use the screen specify the sequence of presenting data and when, in the conversation with the reservations clerk, the system will use the screen. Also, access controls are a feature of the operating system; they are not specified in the DBMS language.

Producing Reports

Reports are hardcopy displays of information produced by information systems. They include one-page summaries, voluminous records of details, and such action documents as paychecks. Team members may specify the production of reports using stand-alone report writers, report writers within fourth generation languages, or problem solution languages. Data developed as an intermediate part of a process or data filed in a database may be the source for a report. If intermediate data is used or the report is an action document, the module that does the processing may produce the report.

The description of report writing here applies to modules dedicated to producing reports or to reports produced from the program modules described above.

The team members describe the report contents, information sources, sequence, totaling, frequency of preparation, use of special forms, and distribution in the prototype definition phase. The data dictionary describes the characteristics of the information elements that will appear on the reports. All of this information is a source for producing the report for the initial model. The only aspect to be decided in this phase is the report's format.

If a report is not to be built completely for the initial model, you only need to produce an identification or a heading for it.

The user members of the project team will play a key role in laying out the reports. If a freestanding or a fourth generation language report writer that is easy to use is the tool chosen for report writing, user members should also be able to do the report writing. If the report writer is not so easy to use or a problem solution language is used, programmer-analyst team members will have to write the reports.

Examples of differences based on tools The prototyping project team developing a batch payroll system decides to produce complete checks and check stubs using COBOL. Their layout is based on the design of the new special form for checks and stubs. The input will be the program-to-program file described in an earlier example. User members of the team will lay out the checks and check stubs. A programmer-analyst will write the program, because COBOL is being used.

An on-line casualty insurance claims system will produce a letter to be sent to reinsurers to notify them of a claim against an insurance policy for which they are providing reinsurance. Preliminary information about the claim from the insurance database will be a source for the report. A fourth generation language containing a good report writer is being used for this project. A claims department supervisor on the project team with training in the report writer can lay out the report as he writes it.

The team developing a prototype based on a new financial system package will produce a report of income and expense by project for a specified time period from the database maintained by the package. Project code is an element the organization is adding to the system; it is not part of the package's elements. Instead of using the report writer provided with the package, the project team is using the stand-alone report writer that is the organization's standard. A supervisor on the team from corporate accounting will lay out the report based on the report description and the data dictionary. The stand-alone report writer is somewhat easy to use. With assistance from time to time from a programmer-analyst, the supervisor can also produce the report.

The team changing a customer profile system into an airlines reservations control system will produce a report of passenger trips taken and the fares received for them for a specific period of time. The team will use the report writer which is a part of the microcomputer relational DBMS language in which the system will be written. A manager from accounting will lay out the report based on the description of it in the prototype definition and based on the data dictionary. Even though the report writer is fairly easy to use, a programmer-analyst will write the statements to produce the report.

Communicating with Other Systems

You should consider the prototype's communication with other systems as you build it. The prototype probably communicates with other systems in at least one of the ways listed in Table 5-14.

If the system will process data from another system, the team should secure a test copy of the file and include it in the initial model of the prototype.

TABLE 5-14 Communication with Other Systems

- Processes data from other system
- Shares database with other system
- Produces data for other system
- Uses data converted from other system

As described earlier under Building the Database, if the prototype will share a database with another system, the team should have a test version of the database built for the initial model.

If the system being prototyped will produce data that will be used by another system, the team should produce that data from the initial model. The team should arrange to have the data the prototype produces run through the receiving system's test system each time they change it significantly.

If the system being prototyped will replace an existing system and if data from the existing system will be converted to the database of the new system by a computerized process, the conversion process itself should become part of the prototype. The system flow describes the conversion. The team does not have to build it as a part of the initial model. But, since it is a source of test data for the database, implementing it for the initial model may be a practical thing to do.

Version Control

The section above on Choosing the Tools briefly discusses tools for maintaining version control. Theoretically the team will not establish version control until after the initial model has been built. Actually, during the building phase, problems may arise that will cause the team to develop additional versions of program modules, files, or a database. If such a need arises, refer to Chapter 6 for a more thorough discussion of this subject.

Delaying Controls and Validations

The prototype definition should describe any controls planned for the system. But, the initial model should not include controls nor validations. The team should implement them in the latter part of the prototype exercising phase.

SUMMARY

1. Building the initial model, which is a software mock-up, includes building:
   - Database
   - Program modules
   - Screens
   - Reports
   - Program-to-program files
2. Tools that help you prototype:
   - Data dictionary
   - Interactive testing
   - Version control
   - Fourth generation language
   - Relational DBMS
3. Differences are allowed between the initial model and the final version of the prototype for:
   - Media of database or program-to-program files
   - DBMS
   - Programming language and command language
   - Specific input or output devices or forms
   - Computer
4. Possible advantages of building a prototype on a microcomputer first:
   - Easy to use
   - Start up quickly
   - Easy to modify
   - Available
   - Accessible
5. Definition of database: A database is a file used by more than one system and accessible in more than one way.

6. Definition of file: A file is a collection of one or more record types.
7. When building the data base use access modules to insulate the using programs from the actual DBMS.
8. Program modules must move data by reading their inputs and producing their outputs.
9. The tools available for screen painting and report writing determine which team members can do them.
10. Users should lead the laying out of screens and reports.
11. Consider the prototype's communication with other systems as you build it.

Chapter Six

EXERCISING THE PROTOTYPE

This chapter describes exercising the prototype, the different stages of exercising it, what you need to exercise it, and how to do it. The chapter also discusses expanding the project team, planning, and progress monitoring.

Chapter 4 describes defining the prototype and Chapter 5 describes building the initial model of the prototype from the definition. This chapter describes exercising the initial model built according to the guidelines in Chapter 5.

Exercising a prototype is testing it, training with it, modifying it, and expanding it until it becomes the system. You begin the exercising process with the initial model; you finish the process by producing the system.

EXPANDING THE PROJECT TEAM

Exercising the prototype requires modifying and expanding the same components of the prototype that were built as part of the initial model. The same skills are needed to exercise as were needed to build, depending also on whether a fourth generation language is used. Table 5-4 summarizes the components and who should work on each. Insofar as possible, the same team members who worked on building the initial model should stay with the project to exercise the prototype.

Exercising the prototype requires two additional tasks:

1. Maintaining version control.
2. Developing draft user guide.

Maintaining version control must be the primary responsibility of one member of the team from Information Systems, designated as the exercising coordinator. If the project is large enough, Information Systems may assign one or more additional persons to help. Those chosen for this task should be junior programmers with clerical skills or documentation specialists. The exercising coordinator should be responsible for version control and integration, possibly along with some programming or documenting. If the project's size does not justify adding another person to the team, one of the programmer-analysts must assume this responsibility.

Developing the draft user guide should be the primary responsibility of one or more technical writers or documentation specialists. If the project is small, the same person from Information Systems may develop the draft user guide and act as the exercising coordinator.

STAGES OF DISCOVERY

The prototype exercising sessions, which may be called simply the "prototyping sessions," work best if they are done in three stages. Table 6-1 lists the focus of each stage of discovery.

The focus of the prototyping in the first stage of discovery is the *data*. How do the reports and screens look to the users? Is the database adequate? Are the interfaces with other systems right? Individual sessions should concentrate on using screens. Other sessions should be for reviewing reports. The prototype should be run to ensure it can read the input from feeding systems. Also, the test versions of systems that use data from the prototype should be run to make sure the prototype provides useable data. The team should make all necessary modifications to the screens, reports, and interfacing files. They should also make any changes required to the database and program-to-program files.

After reports, screens, database, and interfaces settle down, the team should focus next on the prototype's logical and mathematical processes. This is when the team should replace any languages temporarily used for processing with the language to be used in the final version of the system. Where processing logic has been dummied

TABLE 6-1 Exercising the Prototype

Stages of Discovery
1. Data
2. Processes
3. Edits, Validations, and Controls

or done in broad outlines only, the team must gradually flesh out the full logic of the system. Similarly, the team members must build the full mathematical processes needed for calculations. Building these processes may also require the team to make changes to the database and program-to-program files.

The last stage of discovery for prototyping is *validations, edits,* and *controls*. By the time this stage begins, the prototype should be relatively stable. The users should have gained considerable experience in using the system. This experience is important for determining what validations, edits, and controls to build into the system. Gaining this experience first is important, because the cost of developing validations, edits, and controls may constitute more than 50% of the total development cost of a project. With a feeling for how the prototype works and how much additional work and cost any excessive validations, edits, and controls may necessitate, users often trim back what they would have otherwise asked.

RESULTS OF EXERCISING

Defining the prototype produces a definition. Building the prototype produces an initial model. Exercising the prototype produces the results listed in Table 6-2.

TABLE 6-2 Results of Exercising the Prototype

- The System
- Test System
- User Documentation
- Requirements Agreement
- Cadre of Trained Users

The System

Prototyping is the means the team uses to design the system. Furthermore, since the prototype becomes the system, prototyping is the means the team uses to develop the fully operational system.

When prototyping is completed the conversion process has been tested; the screens, reports, program-to-program files, and the database are finished; the processes are all working; the validations, edits, and controls have been set in place; and materials have been prepared in draft form for the user guide and help screens. The system is ready to be installed.

Test System

A test system consists of test data sets, tools, and controls for the controlled and comprehensive testing of an information system or a prototype of it. Comprehensive testing of a system is testing of all its features. Controlled testing is planned testing done without unnecessary duplication of effort.

Testing, when done as a part of a non-prototyping project, is often divided into three categories:

1. unit testing
2. system testing
3. acceptance testing

Unit testing is testing an individual component of the system to verify that it functions according to specifications. The person, often a programmer, who produces or modifies a component usually does the unit testing of it.

System testing is testing all parts of the system together and testing the system's communication with other systems. Information Systems usually does system testing to verify that the system operates as intended.

Users do acceptance testing to determine whether the system meets their requirements and is acceptable to them. This is usually a final stage before a system is placed in production.

Prototyping allows the team members to perform all three types of testing as they exercise the prototype.

You should develop the test system to help you exercise the prototype. When this phase is completed, you should keep the test system to help you implement the system and test later changes to the system.

You can do good testing without special tools, but you cannot do good testing without a plan, thorough test data sets, and adequate controls.

Reasons for testing Testing satisfies a variety of needs. Some are more important when a system is first developed. Others apply especially when modifying and maintaining a system. Table 6-3 lists the principal reasons for testing.

When you are prototyping, testing is a means of designing a system. Testing gives the project team experience with the system, enabling them to make informed changes to the design. The project team continues testing and modifying until they produce a satisfactory design.

TABLE 6-3 Reasons for Testing

- Designing
- Making sure features function correctly
- Ensuring nothing was changed inadvertantly
- Determining whether system will fail
- Training users and operators
- Determining resources required
- Researching problems

A basic purpose for testing is to ensure that running a system or modifications made to it produce predictable results. If a feature of a payroll system is the automatic recalculation of an employee's base pay after notice of a pay increase is received, testing discloses whether the feature works as expected.

Another valuable use of testing, which is sometimes overlooked, is ensuring that adding or changing a system feature does not inadvertantly change something else in the system that was not supposed to have been changed.

Testing may also disclose whether a system may be made to fail. The design criteria for the system should include allowable tolerances for bad data or bad operating procedures. Can bad data make the system fail? Can an operator's loading the wrong version of a file go unnoticed? Testing should provide the answers to these questions.

The testing done as a part of prototyping also trains at least those users on the project team in using the system. As they complete the system they learn how it works.

Testing may determine the computer time required to operate a system and the response time for system functions. Testing may also verify that sufficient storage has been allocated for the system.

Another use for testing, when a system is failing, is researching what may be causing the failure. With testing, you may, for example, isolate the area within the system where the failure is occurring.

Test data sets The heart of any test system is the test data sets it contains. Table 6-4 lists the contents of a test data set. Basically, a test data set consists of a description of what the set will test, database contents, inputs to process against the database, command language statements to run the test, and the expected results. All components of each test data set should have the same version identifier to link them together and to distinguish them from the components of other test

TABLE 6-4 Test Data Set Contents

- Description
- Database contents
- Inputs
- Benchmark outputs
- Expected results
- Command language statements

TABLE 6-5 Test Data Set Log

- Version number
- Description
- Date prepared/modified

data sets. For example, all the parts of one test data set may be labeled as version "A" and the parts of another for the same system may be labeled as version "B." The records kept on test data sets should contain such information as is listed in Table 6-5.

The database contents for a test data set consist of the records the database should contain before a test is run with it. The database contents may be one of four different types. Table 6-6 lists the types of data sets.

Volume data sets contain a large number of records. They are often copies of production databases. A copy of the payroll-personnel database is an example.

Selected data sets contain a relatively small number of records which represent a wide spectrum of data conditions. The ideal selected data set allows full testing of a system's features with a small amount of data. Selected payroll input transactions containing a few samples of each different transaction type and containing data to test all the dif-

TABLE 6-6 Data Set Types

- Volume
- Selected
- Special
- Production

ferent combinations of conditions that the payroll system should handle are an example.

Special data sets contain records for testing a particular feature of or modification to a system. After the feature or modification is incorporated into the system, the team should add the special data set to a selected data set for the system. If a new pay type code is added to the payroll system, a special data set to test the combinations of conditions that may occur in processing transactions with the new code would be a special data set.

Production data sets are produced or maintained by an operational information system. The computer operations organization often controls them. The actual payroll-personnel database is an example of a production data set.

If a system operates differently during different processing cycles, producing different outputs for example, you probably need a different selected test data set for each processing cycle. Examples of processing cycles are daily, weekly, monthly, and quarterly. You may also want separate volume test data sets for at least some of the different processing cycles. Processing cycle test data sets may not all contain inputs, or database contents, or outputs. As an example, the test data set for the monthly processing cycle may consist of monthly benchmark reports and the command language statements to run the system to produce them.

The inputs for a test data set consist of input transactions to be run through the system. Inputs will be one of the four types of data sets listed in Table 6-6. Insofar as possible, they will be machine-readable. If they are transactions for on-line entry and a data entry simulator is available, they will be on a file. Otherwise, they may be on paper for use by the persons who will key them into the system.

The expected results for a test data set consist of database results and benchmark outputs.

The expected results for the database may include machine-readable expected contents. They would depict what the database should contain after a test is run using the test data set. If you have a file comparison utility, you may use it to compare the test results with the contents of the test data set database. Otherwise, the expected results for the database may be a listing of the expected differences in the contents of the database after the test has been run, which you may manually compare to the database contents.

The database contents may be one of the four different types listed in Table 6-6. If the database contents are a special data set, before the test may be run, you may have to use a utility to add them to the database, which may contain a selected data set.

The benchmark outputs for the machine-readable files, reports, and screens depict what they should contain after the inputs are run against the database for the test data set. They should be on paper or in machine-readable form depending on whether you will use a file comparison utility to review them.

The command language statements for a data set consist of the statements required for utilities to load and unload data sets and the database, statements to use data entry simulators and file comparison utilities, and the statements to run the prototype.

Tools for testing Depending on the characteristics of a system and the manner in which you test it, testing tools may significantly increase the effectiveness and efficiency of your testing. But, no matter how helpful they are, no tool can make up for a lack of planning, having good test data sets, or using sufficient controls. Table 6-7 lists the principal types of testing tools.

Interactive testing enables you to submit tests immediately in a computer use environment where they will be run without much delay. Interactive testing allows you to test on-line features as the test is running and to sample batch results on-line without waiting for paper copies to be printed.

A *file loading utility* loads a test system's database and input files from the test data set to be used for the test. You may also use it after a test has been run to copy the contents of the database and output files to other media or even to print them.

A *file comparison utility* compares two files, highlighting any differences between them.

A *data entry simulator* allows you to test the prototype with a file containing transactions you would have entered from a terminal. It processes the transactions in batch mode or simulated on-line fashion, and it produces the results for any output screens as a file, a report, or a screen.

Testing controls The principal tools for controlling testing, as shown in Table 6-8, are test logs and version controls of programs and data. The purpose of controlling tests is to keep the prototype running, to avoid unnecessary effort, to make repeating tests possible, and to keep a record of what was tested.

TABLE 6-7 Tools for Testing

- Interactive testing
- Data entry simulator
- File loading utility
- File comparison utility

TABLE 6-8 Testing Controls

- Version control
  Programs
  Test data sets
- Test logs

Adding data to a database or lines to a program sometimes causes the system to fail. When this occurs, rolling the status of the database and the program back to their condition before the failure will allow prototyping to continue. You may later re-enter the changes to the data or the program to help isolate and correct the problem.

Maintaining a centralized record of what has been tested helps team members avoid duplicating tests done by others. Since they are often not inclined to maintain tight control and keep good records of even their own tests, because of the work involved, this control may even help them avoid duplicating their own tests. Centralized record keeping and control also make it easier to repeat tests that were made earlier. When an earlier test revealed a problem which has been corrected, the test should be repeated to ensure that the correction works.

The record of what tests were run and when they were run is also valuable for determining that the prototype was adequately exercised and tested.

Duties of the exercising coordinator The responsibility for controlling tests and determining which versions of the prototype's components will be used should be the specific assignment of one of the team members, designated as the exercising coordinator. Table 6-9 summarizes this person's duties. If the work warrants it, one or more additional team members may assist the exercising coordinator. Maintaining this control should not be left to individual team members to do incidentally to their exercising the prototype.

TABLE 6-9 Duties of Exercising Coordinator

- Responsible for control
- Copy before changing
- Back out changes that fail
- Record tests
- Maintain sufficient versions
- Maintain test data sets
- Test for inadvertant errors

No changes should be made to a module, screen, report, database, or test data set without the exercising coordinator's knowledge and not before the exercising coordinator has copied the existing version of the component to be changed. This will ensure that the previous version, which probably worked, may be used if the new version does not work.

Before changing to a new version of a module, the exercising coordinator should test the system with the current selected test data set to ensure that nothing was changed that should not have been changed. If he discovers an inadvertant change, he should refer it to the project leader and suspend changing to the new version of the module that caused the problem.

The exercising coordinator should maintain test data sets. If a member of the project team asks that data be added to a test data set, the exercising coordinator should run it through the system to produce all the new outputs. Whenever data is added to a database the exercising coordinator should run the prototype with the database to ensure that it does not cause any component of the system to fail. When data is only temporarily added to a database, he should determine whether equivalent data will be added to the database for a final test of the prototype.

If a change of any kind fails, the exercising coordinator should be able to back out the version of the prototype's components that caused the failure and replace them with the prior version that worked. Not only may changing a program, screen, or database cause the prototype to fail, but also processing new data sometimes causes it to fail. This is especially difficult to control when the data is entered through screens into an on-line system. If such a failure does occur, the exercising coordinator should return the prototype to its status before the data that caused the failure began to be entered. Eliminate the data which caused the problem from the prototype's input until you learn why it causes the system to fail.

The exercising coordinator should keep an accurate record of which version of the prototype's components and test data sets were used for each test.

The exercising coordinator should also maintain sufficient versions of the prototype's components so that the project team may keep the prototype running and repeat any tests the members wish to repeat. If a lack of storage necessitates eliminating older versions, the exercising coordinator should make sure, before deleting a previous version, that a later version of a system component does not cause the system to fail.

Version control The basis for version control of the prototype's components is a version control log. Each active version of a compo-

TABLE 6-10 Version Control Log

- Component identification
- Version identification
- Library identification
- Date and time stored in library
- Description
- Who prepared it

nent should appear on this log. Table 6-10 lists the contents of a version control log. Each individual component should contain its unique identification and its version identification. For example, if a component is a database it should have the identifier given to the database, an indication whether it is the initial or resulting contents, and the version identifier that distinguishes it from other versions of the database contents. If the component is a program module, it should have the module identifier and the version identifier.

The component identification on the version control log tells what the component is (for example a program or a test input), the identifier given to the component, and the version identification to be used for this version of it. Related components that each have their own unique identifications may have the same version number. For example, all the parts of each test data set should contain the same version identification.

The library identification tells what library the version of the component has been stored on. The date and time records when it was stored on the library.

The description briefly tells what makes this version of the component different from other versions of it. Who prepared the version identifies the person responsible for it.

Test logs The exercising coordinator should maintain a record of each test on a test log whose principal contents are listed in Table 6-11.

TABLE 6-11 Test Log Contents

- When test began
- When test was completed
- Test data set used
- Purpose of test
- Versions of components used

The log should contain the date and time each test began and when it was completed. It should also contain the identification of the test data set used.

The log does not have to contain the specific version of all components when the last versions that were stored in the library before the test began are the same ones used in a test. But if any version of a component, other than the last one stored in the library, is used, then the exercising coordinator should list the specific identification of that version.

A brief description of the test should accompany the other log entries.

User Documentation

User documentation for a system consists of a description of what the system does, how to enter input into it, and how to make it produce results, such as reports.

The prototype logical definition should be a good basis for this user documentation. But, as the prototyping progresses, the team adds or changes system features and functions, moving the prototype away from the definition. If the team members do not document this in some way, it is practically lost, to be reclaimed only from the team members' memories and the system itself.

Developing user documentation in draft form is an essential part of exercising the prototype. Developing the final version of the user documentation while exercising the prototype would be premature and consequently wasteful. But developing draft documentation from information that is produced as the exercising progresses should be done so the information is not lost.

If they have sufficient time, the team members responsible for developing the draft user documentation may also change the focus of the logical definition into part of the user documentation. They may then modify and add to this as they prototype.

Requirements Agreement

The requirements agreement for an information system is a statement, to which both the user and Information Systems agree, of what the system will do. Table 6-12 lists the subjects that a requirements agreement should include.

TABLE 6-12 Elements of a Requirements Agreement

- Inputs
- Outputs
- Database
- Processing rules
- Validations and Controls
- Service Constraints

The prototype definition (see Chapter 4) is a first draft of the requirements agreement. Building the prototype as described in Chapter 5 makes the definition concrete. Exercising the prototype enables you to complete the design of the system with the user and thus to determine the system requirements.

When exercising the prototype has been successfully completed, Information Systems and the user will have reached a common understanding of the requirements of the system. Some even view this agreement on the system requirements as the principal value of prototyping.

Even though the requirements will not have been neatly documented when the exercising is finished, Information Systems and the user should formally give notice that they agree on them. In the absence of thorough written requirements, they may point to the prototype itself, the data dictionary descriptions of the information in the system, and the draft user documentation as the record of the system requirements.

When the team finishes the documentation of the system in the implementation phase, that documentation (primarily the user guide) will enhance the documentation of this requirements agreement. But not having that documentation in this phase should in no way lessen the fact that the agreement was made when exercising the prototype was completed.

Cadre of Trained Users

To exercise the prototype the team members run it. At first they work with the initial model. As they gain experience with the system they change the prototype, then work with the changed version. When exercising the prototype is finished, the team members who have been working with it will have learned how to use the prototype. The prototype becomes the system. Those users on the team constitute a cadre trained in the system who can help in the implementation phase that follows.

Requirements for Exercising Successfully

Several factors are necessary to make exercising the prototype successful. Table 6-13 lists them. First and foremost is securing the involvement of the users in exercising the prototype. The users on the team may not have to produce anything; they may not write reports or paint screens. But they do have to participate in the prototyping sessions and in the team reviews of what happens in sessions they do not personally engage in; they have to work with the prototype and to say what they want and don't want. Also, users who are not actively working on the team, especially managers, should participate in the team reviews. Although user involvement does not guarantee the project will succeed, lack of it may cause the project to fail.

Version control, described earlier, enables the team to change the prototype often and quickly without creating confusion.

A test system, also described earlier, allows the team to test the prototype frequently, expanding what is tested, while minimizing the work required to run the tests and to review their results.

Exercising the prototype frequently develops and maintains the team members' familiarity with the system. This enables them to learn the system and to modify it without having to have extensive documentation. If too much time elapses between exercising sessions, the participants may forget how the system works. This may only harm the project. Team members should take part in an exercising session every working day, or at the most every two days.

TABLE 6-13 Factors to Make Exercising Successful

- User involvement
- Version control
- Test system
- Exercising frequently
- Changing quickly
- Communicating changes
- Project management

Changing the prototype quickly helps you to exercise it frequently. Different prototyping sessions may focus on different features of the system, so any particular session may not depend on what happened in the last one. But you should exercise a modification soon after the team decided to make it. Using tools like fourth generation languages often makes changing the system quicker and easier than not using them. If the tools being used or the scope of what is to be changed requires excessive time for making changes, break the changes up. Instead of taking two weeks to make one large change, take three days each to make three small ones.

The prototype contents and architecture may not be stable. Make sure that all team members are told about changes in the prototype. They should also learn to check the current status of any system component that they will be working with.

Good project management is essential to keep the exercising work focused. The project leader should ensure that a schedule and plan are prepared for exercising the prototype and that the schedule is adhered to.

HOW TO EXERCISE THE PROTOTYPE

Exercising the prototype proceeds according to the stages of discovery listed earlier in Table 6-1. First you will concentrate on the data, then you will move on to the processes, and, finally, you will cover the validations, edits, and controls.

The materials produced so far in the project affect the perspective you have of the prototype. Available are the problem statement and the study of the present system, both of which describe the functions, the prototype definition and the data dictionary, which describe the components. Each is a departure point for exercising, and each affects your approach to exercising.

Functions discovered when the existing payroll system was studied may include preparing the pay for a group of employees. The prototype definition may include a description of the paycheck and the check stub. The data dictionary may include the formula for calculating the amount of net pay. If you only concentrate on the components identified in the prototype definition, such as the check and check stub, you may not fully experience how the prototype performs the functions it is to perform, such as preparing the pay for the weekly paid employees. If you only concentrate on the functions, you may miss trying some of the components of the system.

Table 6-14 lists the approach that is recommended for exercising the prototype. Basically, you should plan what you will do and develop

TABLE 6-14 Approach for Exercising the Prototype

- Develop initial plan
- Develop initial test system
- Conduct sessions
- Modify prototype
- Complete test system
- Agree on requirements
- Complete system

a test system to aid in doing it. Then you should exercise and modify the prototype until you complete the prototype and those things, such as the requirements agreement and the test system, that go with it.

Before beginning to exercise the prototype the team members, under the direction of the project leader, will develop a plan and a schedule for exercising. The plan and schedule will detail the features of the prototype that should be exercised, who will exercise it, what materials they will use, and when they will do it. Definite milestones are important to this phase as they are to the other prototyping phases.

Develop Plan and Schedule

Table 6-15 lists the primary tasks to be scheduled for the exercising phase.

Developing the initial test system is producing an initial test data set and any utility runs or procedures required for testing the prototype.

A short period of unstructured familiarization allows the team members to experiment with the prototype. Limit the total time dedic-

TABLE 6-15 Exercising Tasks to Be Scheduled

- Develop initial test system
- Unstructured familiarization
- Exercising sessions for each stage of discovery
- Complete system
- Complete test system
- Agree on requirements

ated to this unstructured familiarization to no more than one or two weeks. Also, allow time in the schedule to make any modifications that the team members decide on during this familiarization.

The exercising sessions will focus first on the data, then the processes, and finally the edits, validations, and controls. The approach for planning these sessions is rather similar. For the data phase and the validations, edits, and controls phase use the prototype definition as a source. For the processes phase also add the study of the present system as a source. From these sources list the system components or features that you will exercise. Then, for each component or feature list the elements of an exercising session shown in Table 6-16. Base the schedule on completing these things. Add to it the materials the team members will need, especially which test data sets they should use.

Completing the system is removing all temporary changes that were made to the prototype, such as the temporary use of a relational DBMS instead of the final hierarchical DBMS.

Completing the test system is determining what different selected and volume test data sets will be needed during exercising, consolidating the various test data developed for the exercising sessions into these test data sets, and labeling and logging the materials so they will be useable.

Scheduling the agreement on the requirements and the completion of the system is primarily scheduling milestone dates by which they must be accomplished.

Because the material in this chapter must concentrate one by one on a succession of topics, it may give the impression that the prototype and the system that evolves from it change little from the prototype definition. It may seem as though the definition process identifies and specifies all the data sets and processes that will comprise the system. Actually, exercising may lead the team to change some parts of the system, to eliminate others, and to add still others. Inputs may require additional information elements, the sequence of reports may change, one database record may combine with another, the need for a new feature may arise, and some reports may become unnecessary.

TABLE 6-16 Elements of an Exercising Session

- Develop test data
- Exercise the component or feature
- Review results
- Modify the prototype
- Test the modification

While planning and scheduling the exercising, the team members must especially estimate how many additional components they will discover a need for as they prototype and plan for them, as if they were already known. They should also estimate how many of the known components will be eliminated without being exercised, since estimates for them may be subtracted from the total estimates.

Conversion Process

If a conversion process has been defined for the system, the team members should prototype it the same as they prototype the features that will be permanent parts of the system. The exercising sessions for it should focus on the same three aspects: data, processes, and edits, validations, and controls. Nonetheless, the conversion process itself may not use significant edits, validations, and controls.

Develop Test System

When the team builds the prototype, as described in Chapter 5, they develop some initial test data for the database and for the inputs to the prototype. This data should be adequate for an initial test data set. It should primarily be data for the database and for inputs from other feeding systems. The project team will use it for the unstructured familiarization with the prototype. If it constitutes a good selected test data set, that may help the familiarization. But, it is not essential that it be so, and the team should not expend any significant amount of time to make it a good selected test data set. But, if the test data set that is available will not allow you to move data through all parts of the system, you should augment it so that you can do so.

More important is preparing any utility operations or procedures that may be required to operate the test system. Because of the installation's procedures for leaving data on the media that will be used to exercise the prototype, the team may have to use file loading utilities to load at least some of the data sets before any prototype exercising may be done. For example, they may have to load disk work files from tape files on which the data sets for testing are kept. Utility operations may also be needed to unload the test data sets back onto another medium, such as tape, at the end of an exercising session. The complexity of the prototype, the tools to be used for testing, or the organization's requirements for running a test may also necessitate writing some procedures to help it run smoothly.

In any particular exercising session, especially for an on-line system, the team members may enter some new data from terminals. But, they may also need to put some new data temporarily on the database or into the inputs from outside systems. Another part of the test system to be developed is any utility operations or procedures that allow such additional data to be added to a test data set. The exercising coordinator should also have procedures for maintaining a record of the use of this data. You should not have to add these utility operations and procedures before beginning the familiarization with the system, since it should not depend on their availability.

The team may also want to set up utility operations and procedures for displaying or printing the results of exercising sessions. Doing this should also not delay beginning the familiarization with the system.

Familiarization

A relatively informal period of exercising the prototype should quickly give the team members a feeling of familiarity with the prototype. This should enhance their ability to exercise it and should increase their identification with the system.

To do this familiarization you need a test data set containing data for the database and system inputs and any file load utility operations necessary for loading data sets (see above). The team should run the input through the system and produce all the outputs, reports, and screens that are ready. The team should review these for usefulness.

The team members may then experiment with entering input into the system, whether it is on-line or batch.

This familiarization may disclose inconsistencies between inputs and the database or between the database and outputs. Glaring problems with outputs, such as an unuseable sequence for a report, may become obvious. The manner in which data is to be entered may be awkward or updating the database from certain inputs may not work.

The team members may modify the plan for exercising the prototype because of what they learn during this familiarization period. If they have time, they may also fix obvious problems. As they experiment with they system they may produce some test data which they want to keep.

The exercising coordinator must control any changes the team members want to have made to components of the prototype or to its test data sets.

Training

The users whose work associates them with a feature of the prototype should usually be responsible for exercising the feature. As they exercise it, possibly modifying and exercising it again, they are training themselves to use it. They and the Information Systems people working with them will later become the trainers for the other users of the system.

As they notice different ways to use the features they are exercising, they should ask the persons responsible for documentation to make notes about them for the documentation. They should also make notes for later use when they are training other persons.

Including additional users and Information Systems persons in the direct work of exercising, even though they may not develop the test data, review the results, or modify the system, increases the training the team members will receive. The additional team members should actually work with the prototype, whenever possible. Besides the increased familiarity with the system this will promote, it encourages them to give their valuable reactions to the feature that is being prototyped.

Prototyping Sessions

In the prototype exercising sessions the team members test the prototype, train with it, modify it, and expand it until it becomes the system. The sessions focus first on the data, next on the processes, and finally on the edits, validations, and controls. Table 6-16 lists the elements of an exercising session.

Develop test data The team may have to prepare test data before beginning an exercising session. If test data is needed, the project leader should assign one or more team members to develop it. If the data is for on-line entry, those developing it may only need to gather source documents from which the data may be entered during the prototyping session. If the testing to be done during the session requires that additional data already be on the database or system inputs, they must put it in machine-readable form and request the exercising coordinator to add it to the test data sets. If it should be made a permanent part of a test data set, they should also notify the exercising coordinator of this.

Data exercising The prototype exercising sessions which focus on the data concentrate on how the reports and screens look to the users, whether the database is adequate, and whether the interfaces with other systems work. Individual sessions should concentrate on using screens, reviewing reports, and running the prototype to ensure it can communicate with other systems. The team should make all modifications decided on during the exercising to the database, program-to-program files, screens, reports, and interfacing files.

Data exercising on-line system As an example, a user and a programmer-analyst will exercise an input screen for a new automobile insurance claims system. The user records information about a new claim using the screen. The contents of the screen are listed in Table 5-13 in Chapter 5. Before the exercising session, the user collects five samples of new claims to use. Also, before the session the exercising coordinator copies the database and the program that presents the screen and updates the database from it

The screen was already painted when the prototype was built. It contains a notice of verification of an insurance policy covering the loss, but this is dummied out. Checking whether adequate coverage exists has not been programmed yet.

The user does not like the way he has to enter claimant name and address on the screen. The user also notices that the screen should contain a claim type element. He and the programmer-analyst discover that claim type also does not appear on the database. They interrupt their exercising session to discuss this with the project leader. He approves their modifying the screen and the database.

The programmer-analyst updates the data dictionary. The user and the programmer-analyst modify the screen layout. They inform the exercising coordinator of the change to the database structure. He copies it, then the programmer-analyst adds the claim type element to the data dictionary and to the database structure. The user then enters data for five more claims. He is satisfied with the new screen and the new element.

The user and the programmer-analyst print copies of the new screen with data for later review with the whole team. The user makes a note to pass on to those who will exercise the verification of coverage process that claim amount should be checked against coverage amount based on claim type. They notify the exercising coordinator of the change to the screen. They also request that the data entered onto the database be kept, because it is to be used later to exercise a screen for assigning claims to claim supervisors. The exercising coordinator assigns identifications to the new version of the screen program and the modified database. He first tests the new version of the screen, using the prior test data set, to ascertain that it changes nothing inadvertantly.

Data exercising package modification As another example, a user and a programmer analyst exercise an input screen used for manual journal entries to a financial system package. The screen was painted with a map editor when the prototype was built. The information elements to appear on the screen were mostly predetermined by the package designers, but their size and characteristics were decided when the data dictionary for the package was established. The team also added a non-standard project code to the input screen for budget and expense entries.

The user collects several journal entries from recent journals to use as test data. Before the session begins the exercising coordinator copies the database and the program that updates the screen.

In entering the data the user discovers that account number does not include division code, which appears on all entries other than home office entries. This was not noticed before, because only home office entries were used during the familiarization period.

At the end of the exercising session the user and the programmer-analyst tell the exercising coordinator not to save the updated database. They must discuss the lack of provision for division code with the entire group.

After this discussion, the user adds division code to the data dictionary for the package as a subelement of account number, whose size is expanded. The structure of the database and all inputs, outputs, screens, and reports that use account number are automatically updated by the package's data dictionary.

The original user and programmer-analyst then exercise the journal entry input screen again, after the exercising coordinator has first saved the database contents and the screen processing program. The user and the programmer-analyst are satisfied with the screen this time. At the end of the session, they keep a print-out of the input screens to review with the entire team. With the prior approval of the project leader, they ask the exercising coordinator to save the augmented database contents because they include records with division code as well as some project codes.

Data exercising batch system An additional example is a new batch payroll system. A user and a programmer analyst exercise the printing of the paycheck and the check stub. The source for printing them is a program-to-program file that is being dummied out by the fourth generation language skeleton of the module that will do the payroll calculation. The data that is being dummied out is not rich enough for their exercising session. So, the user has developed another test data set that tests all the elements and features of the paycheck and check stub. The programmer sets up a utility run to load a temporary internal file used by the pay calculation module to produce the program-to-program file with they data they have developed. The exercising coordinator copies the program before they make this change.

They run the paycheck and check stub, but problems with the data they have created keep them from being satisfied with the run. They have to modify the data they put in the calculation module twice before it produces output that gives them a satisfactory run.

They copy these results to review with the team. They ask the exercising coordinator to save the new version of the calculation module and to keep the new version of the paycheck and check stub as partial benchmark outputs. The exercising coordinator tests the modules that use data from the new module, using the current selected test data set, to ensure that the new module does not inadvertantly change anything. The user and the programmer-analyst also provide him a copy of the data and a description of the payroll processing that produced it. The exercising coordinator will label this with the same version number as the data and the output before filing it.

Data exercising system enhancement Another example is a user exercising a screen for making an airline reservation. Figure 6-1 is a sample screen. It is part of a modification to what used to be an airline passenger profile system. The team is changing it into a reservations system. They are using a relational DBMS language to write the modification that will run on a microcomputer.

Before the exercising session begins, the exercising coordinator saves the program that reads the screen. The program that reads the screen does not update the database. Instead, it simply erases whatever data was last on the screen, before it allows the entry of a new reservation.

The user is a supervisor of reservations agents. He has collected several recent reservations containing many different combinations of conditions. He has another reservations agent, acting alternately as a travel agent or a passenger, call him and give him the information over the telephone. As the user enters the reservations, he finds the order in which the screen calls for data to be awkward. It also does not give him the ability to check on seat availability.

At the end of the session the exercising coordinator does not have to save any data and the user has not modified any program.

After the session, the user reviews the results with the other team members and asks for a programmer-analyst to modify the way the screen asks for data. He also requests that a programmer-analyst make the seat availability operation accessible from the screen. The seat availability operation is not new. Only making it accessible from this

```
                         SAMPLE AIR CORPORATION
                        PASSENGER PROFILE SYSTEM
PASSENGER INFORMATION:                                      PNR NUMBER:999901:
                                                            AGENT:KE:
BOOKED BY: Psgr (1), Sec'y (2), or Tr Agnt (3):1:  SALES OFFICE: :
LAST NAME:Lantz           : SAL:Mr   :FIRST AND INIT:Kenneth E    :
HOME STREET:22458 Ventura Blvd.  Ste E    :
CITY AND STATE:Woodland Hills, CA    :ZIP:91364  :PHONE:(818)704-1470:
COMPANY:Atwater Lantz Hunter & Co.    :CORPORATE ACCOUNT NO:    :
TITLE:                              :
OFFICE STREET:22458 Ventura Blvd.  Ste E    :
OFFICE CITY AND STATE:Woodland Hills, CA    :ZIP:91364  :
PHONE:(818)704-1470: SECRETARY: LAST:Milkbucket   :FIRST:Helen     :
SEAT RESERVATION: 5A:  SMOKING?:N: SPECIAL REQUESTS:haircut        :
PASSENGER COMMENTS:Works while travelling; likes good lighting


                                                  :
AMT: 850.80:TYPE: :LIMO: 50.00:PAY FORM:AC :
CR CARD NAME:K E Lantz                  :
CARD/CHECK NUMBER:888888888888888888:EXP DATE:12/86:
TRANSPORTATION TO:FROM AIRPORT:limo:limo:
COPY AND UPDATE (C) UPDATE (U) OR IGNORE (I): :
ENTER h FOR HELP: :
```

FIGURE 6-1 Sample Passenger Input Screen

screen is new. A programmer-analyst makes these changes, still not updating the database. He dummies out the seat availability output. The writer working on the project drafts a description of the use of the seat availability option for a help screen and the user guide.

The user then exercises the screen again. This time he is satisfied with how it works. At the end of the session, he gives a list of the reservations to the exercising coordinator to maintain as an input test data set for later use when the database will be updated. The programmer-analyst also gives the modifications to the exercising coordinator to be made the latest version of the prototype.

Aspects of data exercising Table 6-17 lists the important aspects of prototype exercising focused on the data. All the data identified in the prototype definition provides the basis for the exercising tasks.

Any special test data required for a particular exercising session is prepared before the session. The user often leads the test data preparation. The team members performing an exercising session may ask the exercising coordinator to incorporate the test data they used into a permanent test data set.

The role of the exercising coordinator in controlling the versions of the prototype's components and ensuring no change inadvertantly affects other parts of the prototype is critical to smooth prototyping.

If a project is large and many persons are on the project team, many different prototyping sessions are possible every day. The most

TABLE 6-17 Aspects of Data Exercising

- Exercise everything containing data
- Test data prepared beforehand
- Test data for a session may be incorporated into test data sets
- Exercising coordinator controls all changes to system components and test data
- Many different sessions a day are possible
- Often a user and a programmer-analyst exercise together
- Problems with database sometimes discovered
- Changes affecting other components must be coordinated
- Update documentation
- Results are reviewed frequently with entire team

important limit is how many sessions the exercising coordinator can control.

Often at least one user and one programmer-analyst should be on a team for a prototyping session.

Data exercising, especially since it is the first prototyping done, may disclose problems with the database.

A person with documentation responsibility should record any changes or additions that affect such things as help screens, the user guide, or the data dictionary.

If the team members exercising a particular component discover problems with another component, such as the database, they must review these problems with the project leader or the entire team before making any modifications to the other components.

Team members exercising different components of the prototype should frequently review what they have done with the entire team. These reviews should be at least daily for large projects. No more than two or three working days should elapse between reviews for smaller projects.

Exercising sessions for processes In the second stage of discovery, the team should focus on the prototype's logical and mathematical processes. The team may plan the prototyping sessions for processes from the modules in the physical definition of the system and the processes identified in the logical definition and in the study of the present system. *During these sessions, the team should replace any languages temporarily used for processing with the language to be used in the final version of the system.* They should gradually expand logical processing that initially has been dummied or done in broad outlines only, and they must build the full mathematical processes needed for calculations.

Process exercising on-line system As an example, a user and a programmer-analyst exercise the notification of reinsurers process in a new automobile insurance claims system. This process will determine whether there are reinsurers providing coverage on a policy for which the company has received notice of a claim. If the reinsurer's liability is significant, the process will prepare and print a notice to be sent to them.

The system will be implemented in COBOL on a mainframe, but the prototype has been done in COBOL on a microcomputer. The database will be implemented with a hierarchical DBMS, but it is being simulated on an indexed file access method available with the microcomputer COBOL processor.

The reinsurance notices were originally dummied out. A user enters a request on a screen to print all new notices, since the last time they were prepared. The user, who is a claims department supervisor, guides the programmer-analyst in expanding on the processing described in the logical definition of the prototype and in the study of the present system. The programmer-analyst spends about two full days developing coding for the process. He uses modules developed previously to access the insurance database containing both claims and policy information. He does not need the user to help on the format of the notices, since they were previously dummied out.

While the programmer-analyst is developing the modules, the user reviews the contents of the selected test data set. He develops some additional data to cover fully the conditions that may occur in this processing. He adds this data to the database through the claims notification process and the supervisor assignment process, after first notifying the exercising coordinator. The coordinator saves the database contents.

The user lists the notices that the process should produce. When the programmer-analyst has finished the module, he and the user exercise it. It does not completely produce the results the user expected, partly because of errors in the expected results and partly because the module does not function as the user had hoped. He modifies the expected results and the programmer- analyst modifies the module. This takes another day before they are ready to exercise the process again. This time the results are as the user had hoped.

They ask the exercising coordinator to replace the database contents for the selected data set they were using with the new version they have created. They also ask him to replace the program module for producing the notice with the new module. He will check for inadvertant changes before changing modules. They keep copies of the reinsurance notices to review with the entire team.

Process exercising package modification An example of process prototyping for a package modification is exercising the trial balance produced by a financial package. A user who is a supervisor from the corporate accounting department will do the prototyping. The account table the system uses to define what the accounts are and how entries for them are to be processed has been partially completed and some corresponding journal entries made as part of a test data set. The user will add enough data for two accounts to ensure that more than a page will be printed for each on the account activity report. The exercising coordinator will save the database contents first.

The user will run the trial balance and check the results. If he is satisfied, he will add additional accounts to the account table and to the journal entries and run it again. He will continue this process, working with the exercising coordinator, until he has made an entry in the account table for all the accounts. When he is finished, he will ask the exercising coordinator to save the journal entries as a new test data set.

Process exercising batch system An example of exercising a process is exercising the net pay calculation module in a new batch payroll system. The team members will have already reviewed the preparation of the inputs to the module and the printing of the check and check stub from its results (see Data Exercising Batch System above), when they performed the prototyping sessions for the data.

A user, who is a supervisor from the payroll department, and a programmer-analyst will do the prototyping. First, they must develop the process which, until now, they have only been dummying out. The data dictionary contains processing rules for some elements, such as amount net pay, and the function description portion of the prototype definition contains a brief description of the processing. The study notes on the present system contain a further description of the processing. The user and the programmer-analyst will develop the full processing. They will describe it to the person with documentation responsibility so he may draft it for the user guide.

The module, which until this point is written in a fourth generation language, already contains the outline of the processing. It contains statements for reading the inputs and producing the outputs. The ultimate language the team chose for this module is COBOL. Since temporarily used languages are to be replaced during process exercising, the programmer-analyst, in his first expansion of the module, will redo what was written in the fourth generation language and add the logic but not the mathematical processes.

While the programmer-analyst is doing this, the user gathers test data to add to the selected test data set they will use. One source he will draw from is the data that was used for exercising the check and check stub, as described above. He also lists what he expects the contents of the resulting paychecks and check stubs will be.

Before modifying the selected test data set for them, the exercising coordinator saves the old version. He also saves the existing version of the net pay calculation module.

When they are ready, the user and the programmer-analyst exercise the module. This exercising demonstrates whether the module chooses the proper input records, even though it does not yet actually calculate the amounts.

After they produce satisfactory results with the incomplete module, the programmer-analyst adds the mathematical calculations to the program. Then they perform additional prototyping sessions and make any necessary modifications until the module produces satisfactory results.

The user and programmer-analyst save the results of the successful exercising sessions to review with the entire team. They ask the exercising coordinator to change to the new module, test data set, and benchmark outputs for the check and check stub. Before doing this the exercising coordinator must run the prototype using the old test data set and the new module to check whether anything was inadvertantly changed. If not, he makes the requested changes.

Process exercising system enhancement A final example of process exercising is a user and a programmer-analyst exercising the reservations entry process of an airline reservations system (discussed under Data Exercising System Enhancement above).

The same user has previously exercised the screen, including a dummied version of the seat availability check. The user may use the test data developed for the prior exercising for this process exercising.

The programming language being used is the one provided with the relational DBMS used for this microcomputer-based system. Although the program must access and update five different relational databases, the programming to be done is straightforward. The programmer-analyst asks the exercising coordinator to save the current program that reads the screen before he begins working on the new version.

After the programmer-analyst has completed writing the program, he and the user exercise the new module. They discover a few size discrepancies between elements on the database and on the screen. The database agrees with the data dictionary, but the screen does not. They modify the screen and exercise again until they are satisfied with the results. They save copies of the screen input and selected database output to review with the entire team. They ask the exercising coordinator to make the new version of the screen program the current version. He does after first running the rest of the prototype with the database containing the data temporarily added to it to ensure it does not cause another part of the system to fail.

Additional aspects of process exercising The aspects of data exercising described above also apply to exercising the processes. Table 6-18 lists a few additional aspects that apply to process exercising.

TABLE 6-18 Additional Aspects of Process Exercising

- Change to final programming language now
- First full development of processing requirements
- One exercising assignment may require several mandays

If another programming language has been temporarily used for a module, the team should change to the ultimate language as the process exercising begins. For example, if a fourth generation language was used for a module, but COBOL will be the language for the final version of it, the team should rewrite it in COBOL as they begin the process exercising of it.

The team may never have developed the full requirements for a module before the process exercising of it. They may have studied the processing when studying the present system and they may have identified it in the definition of the prototype, but they may not have developed the full requirements for it.

Because developing the requirements for a process, preparing test data for it, changing programming languages, and exercising it until it is correct may require considerable effort, several mandays may be required to exercise some modules.

Exercising for edits In the third stage of discovery the project team should focus on the edits, validations, and controls for the system. The prototype definition and the data dictionary are the principal sources for doing them.

Edit: changing the value of one or more information elements.

Validation: inspecting data to determine whether it is acceptable.

Controls: procedures and data used to ensure that only correct data is processed.

Often more than half the expense of developing and programming a new system is for its edits, validations, and controls. Many times elaborate ones are developed to prevent something that is logically possible but that might only rarely happen, from happening. An important reason for not focusing on them until the third stage of discovery is to give the users experience with the system they have been helping to prototype before they decide whether to implement them. This experience often leads the users to simplify or to view as unnecessary edits, validations, and controls which they might have otherwise requested. The time that will be required to add them to a system the users are familiar with and enthusiastic to begin using also may influence them to be sparing in their requests.

The team members do not define the edits, validations, and controls when they originally complete the Information Element Descriptions as part of defining the prototype. As they add edits, validations, and controls in this phase, they should update the data dictionary with descriptions of them.

Edit exercising on-line system Exercising an automobile insurance claims system to verify coverage is an example of exercising edits, validations, and controls. Part of the information entered on a new claim, as described under Data Exercising On-line System above, are estimated claim amount and policy number for the insurance policy against which the claim is made. If there is not adequate coverage for that large a claim under the policy, the system should notify the clerk who is entering the claim.

A user, who is a supervisor from the claims department, and a programmer-analyst exercise this verification. The current version of the program checks the policy coverage record on the database and dummies out the notice on coverage. The user and the programmer-analyst develop the procedure for checking claim type against coverage type for a policy and determining whether the policy is still in force. The supervisor adds a reference to this procedure to the validation description of the claim amount element in the data dictionary. The programmer-analyst adds this processing to the module, after the exercising coordinator first saves the old version of the module and the database. The user develops some test data to test the various conditions that may occur.

When the modified module is ready, the user enters the test data he has developed. The program produces the results that the user expected. They save a copy of the results to review with the other members of the team.

The programmer-analyst asks the exercising coordinator to make the modified module the new version and to make the input data part of the selected test data set. After ensuring that the new module does not cause an inadvertant failure when a test data set is run against it, the exercising coordinator makes it the current version for the prototype.

Edit exercising package modification Adding edits, validations, and controls to the journal entry processing of a new financial package is an example of such work with a package modification.

The team wants to add edits to change the value of some obsolete account numbers on entries from feeding systems to the account numbers that are currently used.

They also want to activate the batch control processing which is part of the package, but which they have not yet used. The batch control processing suspends action until corrections are made on out-of-balance batches or on batches whose input batch totals disagree with the batch totals the system computes.

In earlier exercising sessions team members finished making entries in the account table for all the accounts. They now want to activate the option to reject any entries not containing these account numbers. In addition, they want to use options in the account table to reject the entry of any accounts for a department except those specified as valid.

A supervisor from corporate accounting and another accountant along with a programmer-analyst form the team to do this exercising. The users make all the changes to the account table, after first notifiying the exercising coordinator so he may save the current versions of the database and the test data set that they want to augment.

The first entries they make should be the edits for obsolete account numbers. While they are doing that the programmer-analyst should extract test files from the test systems of the feeding systems. When he has this data, he may exercise with it. One of the users should review the results.

As the users make the other account table changes, they should add data to the test data set to check that both valid and invalid entries are properly processed. They should also prepare batches that exercise all the batch control options.

When they have successfully completed the exercising, they should identify for the exercising coordinator the test data set, account table, and database that should become the new versions. They should also save printouts of the results of the exercising session to review with the other members of the team.

Edit exercising system enhancement Exercising edits and validations for the reservations input screen in an airline passenger profile system that is being enhanced to become a reservations system is another example. Figure 6-1 shows a sample screen. The team members are using the microcomputer DBMS language that the profile system was written in to make the changes.

A user who is a reservations department supervisor and a programmer-analyst are doing the exercising. The user wants some simple edits including the automatic capitalization of the first character of each name field and the removal of all commas from certain fields, such as name and address fields. The user wants all fields that should be numeric validated that they are and amount fields validated for maximum amounts. In addition, the user wants the booked by element validated for one of three valid values and the company name validated against the corporate account number assigned to it. The user adds the specification of these validations to the data dictionary.

Before beginning work on the changes, the programmer-analyst asks the exercising coordinator to save the current version of the screen input program and the current version of the database. While the programmer-analyst adds the edits and validations the user prepares test data and expected results with which to exercise the screen.

The user and the programmer-analyst exercise the screen until they are satisfied with the results. They save a copy of them for review with the other members of the team. They ask the exercising coordinator to add the test input to a selected test data set and to use the new screen program as the latest version. Before he does this, the exercising coordinator runs the prior test data sets with the new version of the program to ensure that no inadvertant changes were made.

Aspects of edit exercising The aspects of data exercising described in Table 6-17 above also apply to exercising edits, validations, and controls. Exercising edits, validations, and controls is similar to exercising processes in that one exercising assignment may require several mandays of work to complete.

Team Reviews

Team reviews are reviews, by all the members of the prototyping project team and other responsible users, of the results of individual exercising sessions done by smaller teams. They are the heart of exercising, because they enable users who do not actually participate in a session, as well as the Information Systems members of the team, to review and to comment on the results of the session. This promotes the user involvement which is an essential factor for success in exercising the prototype. It also secures better participation by the Information Systems members.

Team reviews should occur at least every two to three days. They should last for one to two hours. If the group cannot review all the exercising done in the prior two to three day period with such a schedule, then the team reviews should occur more frequently. Having several reviews, each lasting no longer than two hours, is better for group enthusiasm and interest than having fewer reviews that last longer than two hours.

Each participant should carefully review the results of the exercising sessions. The project leader should emphasize that the success of the prototyping is the responsibility of the entire team, not just those members who did the work of any particular exercising session.

One new thing that should be done in team reviews for each subject of an exercising session is deciding whether service constraints should be established for the reports, screens, and other outputs produced by that part of the prototype being exercised. If they are needed, the team should develop them and update the allowable response times in the record descriptions with them.

Complete System

If the computer, devices, data set media, command language, or DBMS for the prototype is different from what was planned for the final version of the system, the team members must change to what was planned. This includes moving from a microcomputer to a mainframe, from a temporary database structure to the permanent structure, from command language statements for testing to the command language statements for operation, and from a temporary terminal to the ultimate terminal that will be used. It includes changing from all the things that you may initially do differently as mentioned in Table 5-9, Guidelines for Building Initial Model. It does not include replacing temporary coding in another language; you should have done that as you exercised the processes.

If you must change more than one kind of thing, wherever possible you should try to change only one kind at a time. Test each change, then make the next change and test it. If, for example, you must change terminal devices and the DBMS, change the terminal devices first. After you have successfully tested that with the smallest possible test data set, then change to the ultimate DBMS and test that. Working this way will make isolating and correcting problems easier.

In the example of the insurance claims system that has been used here, you would need to change from the microcomputer and simulated database that was used to the mainframe computer and the actual database. This would also imply a change in terminals, since the microcomputer itself was acting as the terminal in the prototype. In this case, you will need to change terminals when you change computers. Also, the mainframe does not have a DBMS like the one used on the microcomputer to simulate the ultimate DBMS. You could use an indexed file access method on the mainframe to simulate the database, but this would be additional work in which you might make errors. Because of this, you may decide to change to the ultimate DBMS at the same time as you change computer and terminal. You should test this thoroughly with all the test data sets that you have.

Complete Test System

If the exercising coordinator has not been able to keep up with all the changes and additions that were made to the test data sets during the prototype exercising phase, he should complete them. The project leader should also determine whether the team has produced all the test data sets that were planned. If they have not, he should have some team members complete them. This work may continue into the next phase, although it will have to be finished before the acceptance testing can be completed.

Complete Requirements Agreement

The users and Information Systems should formally give notice that they agree on the requirements and that the prototype meets them. Since prototyping has been the means for developing much of the requirements, doing this may be straightforward. Instead of separate documentation of them, they may point to the prototype itself, the data dictionary descriptions of the information in the system, and the draft user guide as the requirements documentation.

PROGRESS MONITORING

Progress monitoring the work of this phase may be more difficult than progress monitoring the work of other phases, because the exercising itself may lead the team members to add new features to the system or to eliminate features from it. Yet the schedule should contain specific milestones to which the team members should adhere.

If the original planning was good, then the allowances made for adding and eliminating features should be good. The principal concern of the project leader may be making specific changes to the schedule and its milestones while staying within the major milestones for the phase.

PROTOTYPE BECOMES THE SYSTEM

When you have finished exercising the prototype, done anything that was necessary to complete the system, completed the test system, and agreed on the requirements,

THE PROTOTYPE HAS BECOME THE SYSTEM.

SUMMARY

1. Exercising a prototype is testing it, training with it, modifying it, and expanding it until it becomes the system.
2. Persons should join the project team in this phase:
   - To maintain version control
   - To develop draft user guide
3. Prototyping stages of discovery focus on:
   - Data
   - Processes
   - Edits, validations, and controls
4. Results of prototyping:
   - The system
   - Test system
   - User documentation
   - Requirements agreement
   - Cadre of trained users
5. Test system consists of:
   - Test data sets
   - Tools
   - Controls

6. Factors essential for successful prototyping:
   - User involvement
   - Version control
   - Test system
   - Exercising frequently
   - Changing quickly
   - Communicating changes
   - Project management
7. Method for exercising the prototype:
   - Plan
   - Develop test system
   - Familiarize team with prototype
   - Train as you go
   - Conduct sessions
   - Team reviews
   - Complete test system
   - Complete system
   - Complete requirements agreement
8. Parts of a typical exercising session:
   - Develop test data
   - Exercise the component or feature
   - Review results
   - Modify the prototype
   - Test the modification

Chapter Seven

IMPLEMENTING THE SYSTEM

The previous chapters describe defining, building, and exercising the prototype. When exercising is finished, the prototype has become the system.

This chapter describes installing the system that the prototype has become. Implementing includes the final documentation, testing, acceptance, training, converting, and installing needed to make the system operational.

DEVELOP PLAN AND SCHEDULE

Logically, at least, the tasks to be done seem distinct, with the beginning of one dependent on the completion of another. For example, completing the user documentation before beginning the user training so the documentation may be used in the training seems ideal. Considerations like this may tempt you to schedule the different tasks serially. But scheduling serially, justifiable though it may seem, should not be done. It will slow the pace of the project and dampen the enthusiasm of those involved with it.

Schedule this phase so that the different tasks overlap each other as Figure 7-1 depicts schematically. You will still need to consider some

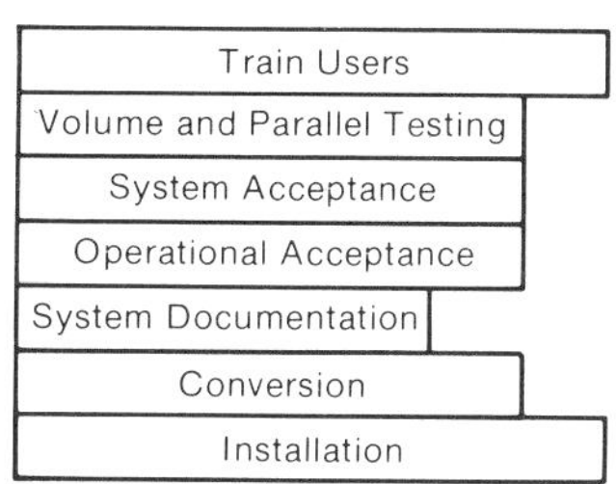

FIGURE 7-1 Overlapping implementation tasks

TABLE 7-1 Task Responsibilities

PRIMARY TASK	RESPONSIBILITY
Train Users	User
Volume Testing	Info. Systems
Parallel Testing	User
System Acceptance	User
User Documentation	User
Computer Operations Documentation	Info. Systems
Conversion	Info. Systems
Installation	Info. Systems

dependencies. For example, you cannot complete the installation until after you have completed the operations documentation. Also, you must complete the conversion, the volume and parallel testing, and the acceptance testing before installing the system.

Although both users and Information Systems persons may work on most tasks, user team members or Information Systems team members may have primary responsibility for different tasks. Table 7-1 depicts this.

RECOMMENDATIONS FOR SUCCESS

To maintain the success of your prototyping project and to ensure a smooth implementation, you should follow the recommendations listed in Table 7-2. Give the team members specific assignments and dates for completing them. Monitor the progress and quality of their work.

Complete the phase quickly. Some of the benefits of prototyping will be lost if the implementation phase drags out. Keep the team intact. This will give you a good number of experienced persons to help you complete the phase quickly.

The team reviews should continue. They will help maintain the team members' enthusiasm and feeling of responsibility for the project. They also provide the team members the opportunity to make suggestions on the work being done by others.

Meeting once a week is a good frequency. Meeting less often may reduce the meetings' benefits. Meeting more often should not be neces-

TABLE 7-2 Recommendations for Success

- Specific assignments
- Definite milestones
- Monitor progress and quality
- Complete phase quickly
- Keep team intact
- Continue team reviews
- Maintain version control

sary. A meeting of one hour duration should be sufficient for the various team members to report on what they are individually doing and to comment on the work that other team members are doing.

Maintaining version control is important for implementing the system, as it was for exercising the prototype. Although the program modules will probably be the same, the team may use different sets of command language statements, database contents, and data sets for training, testing and system acceptance, and conversion and installation.

The exercising coordinator should control the different versions of the system components that are used.

Implementation involves making two copies of the prototype.

- Copy 1 becomes the production version.
- Copy 2 becomes the test system (prototype).

TRAINING REMAINDER OF USERS

All those persons who will use the new system and have not yet been trained should learn why it was developed and how to use it before it is implemented. Having the prototype allows the team to do the training with the prototype and its test files, without waiting for the system to become operational.

Users from the project team with the interest and ability to train should lead the training of the other users. Team members from Information Systems and professional trainers may also assist them. Those training should use any available drafts of the user guide and help screens in the training. Naturally, those team members doing the documentation will want to receive the reactions to it of those who are doing the training.

Using a training room for group presentations as well as practice sessions may be helpful, as long as those being trained can actually work with the prototype, either in the training room or at some other location.

One or more selected test data sets should be useful for the training. The trainers may want to use copies of them, because they may want to keep data that the trainees have entered for use at a later time. If they need to add data to one or more test data sets for training, they should have the exercising coordinator maintain them as additional versions or incorporate the data into already existing versions.

Volume test data sets should not be used for training unless it is absolutely necessary. Using volume test data will significantly increase the computer resources required for training, possibly to the point of seriously restricting the training that may be done during normal business hours.

The training should be completed by the time the system is installed.

VOLUME AND PARALLEL TESTING

Volume testing is done for those facets of the system that are peculiar tto large volume, including the storage allocations for the database and data sets, the command language statements, and the service constraints.

Volume testing of the storage allocations and their associated command language statements are a form of smoke test to ensure the system will handle its expected activity. If problems are discovered, they are often with the storage allocations for program-to-program files rather than with those for the database.

TABLE 7-3 Comparison of Volume and Parallel Testing

VOLUME TESTING	PARALLEL TESTING
• Test storage allocations, command language, service constraints	• Compare to current system
• Volume test data sets	• Production data sets
• No detailed review	• Detailed review
• Information Systems responsible	• User responsible

Volume testing to verify that the system meets the service constraints ensures that the system does not exceed the allowable response times for updating the database, responding to inquiries, or producing reports or other outputs.

Parallel testing is processing the same data through both the new system and the existing system in order to compare the results. Parallel testing is possible if the new system replaces existing processing. A successful parallel test is often a prerequisite for system acceptance. If a conversion is necessary to start up the new system, you may need to do it to obtain the data for parallel testing.

Parallel testing may be real or historical. Either current or prior period production test data sets may be used for parallel testing depending on whether the testing is real or historical. In a real parallel test, while users are waiting for results to do their work, the same data is input to both the new and the old system. Users, who are not on the project team but who are from the parts of the organization that will use the new system, may assist in doing a real parallel test. A real parallel test adds to the users' workload. The additional work coupled with any delays that occur in processing create pressure on those doing the testing. In a historical parallel test, data that was processed for a prior period by the old system is processed by the new system. Any delays that occur in a historical parallel test do not have the importance they have in a real parallel test, because the results are not being used for day-to-day operations.

The team members may make volume test data sets by copying production test data sets or by generating large data sets. Parallel testing is done with production data sets or data converted from them.

The detailed results of a volume test are normally not reviewed carefully; whereas the detailed results of a parallel test are thoroughly compared to the results from running the existing system. The team members may use file comparison utilities, but at times differences in processing methods and results may make the use of such utilities impossible.

Although both users and Information Systems persons may work together on volume and parallel testing, an Information Systems programmer-analyst is likely to lead volume testing and a user is likely to lead parallel testing.

Volume testing is often useful. Parallel testing may be necessary for system acceptance, if it can be done. Doing a real parallel test is not recommended. It creates additional pressure, as described above, and may easily exhaust those who are doing it. Also, if a real parallel test is done, then a volume test should precede it to ensure all possible problems with storage allocations and command language statements have been found and corrected.

If a real parallel test is not done, then a historical parallel test may be combined with a volume test, using the same test data sets and the same runs. Even though only one test will be done, two different groups of team members should be involved, the one using it as a volume test and the other using it as a parallel test. This should ensure that both sets of purposes are satisfied without causing undue delays.

SYSTEM ACCEPTANCE

System acceptance is the verification by the users that the system meets their requirements. It is the formal recognition by the users that the system developed in response to the original service request meets their needs. In addition to what was done during exercising, it also includes the verification that the system meets the design criteria, which cover such things as tolerances for bad data and bad operating procedures. Other than verifying that the system meets the design criteria, system acceptance should be no more than a formality after the users' close participation in the exercising.

Even though the systems acceptance effort should be led by the user members of the project team, Information Systems members should work with them to ensure that they develop adequate criteria for it.

If a parallel test (see above) is possible, users usually perform one to verify the system. Running a successful parallel test assures them that the new system is at least as good as the system it is replacing.

Verifying that the system meets the design criteria probably necessitates additional specific testing. The team members doing it may add additional data to volume or selected test data sets, depending on whether the design criteria to be tested are sensitive to volume. After the verification is completed, the exercising coordinator should add the data that they used to the test system.

The team members working on system acceptance should begin as soon as possible after the initiation of the implementation phase. System acceptance must conclude successfully before the system is installed.

SYSTEM DOCUMENTATION

System documentation describes a system and its history for those who will use it, run it, modify it, or learn about it. Table 7-4 lists the principal components of a system's documentation. As Table 7-4

TABLE 7-4 System Documentation Contents

COMPONENT	PRIMARY RESPONSIBILITY
• User guide	• Users
• Data dictionary	• Information Systems
• Operating documentation	• Information Systems
• Program documentation	• Information Systems
• System history	• Information Systems

shows, team members from Information Systems should be primarily responsible for all but the user guide.

The user guide describes what the system does and how to use it for those who would use it or learn about it. The entries made in the data dictionary for a system describe its data sets and information elements, including their edits and validations and the processing that is unique to them. The operating documentation tells how the system has been installed in a particular hardware and operating system environment and gives instructions for running the system. The programming documentation tells how the system has been programmed, especially as an aid to anyone who needs to modify it. The entries in the system history describe the project for those who want to learn about it.

User Guide

The user guide describes what the system does and how to use it, but, also, along with entries made in the data dictionary, it documents the requirements agreement for the system. The user guide should evolve from the original logical definition of the prototype.

After the prototype has been built and has become the system, maintaining the prototype definition as such satisfies no practical purpose, other than after-the-fact documentation, which most Information Systems persons despise doing. The same is true of the requirements agreement. Yet, maintaining the user guide is necessary for the orderly operation of the system and the training of new staff members to use it.

Table 7-5 lists the contents of the user guide. The logical prototype definition would be the basis for all but the edits and validations, whose source is the data dictionary. If the system is completely batch, its user guide would contain nothing for either screens or help screens.

TABLE 7-5 User Guide Contents

- Report and output screen catalogs
- Functions
- Inputs
- Controls
- Edits and validations
- Help screens

Table 7-6 lists the contents of report and screen catalogs. All the parts of the logical definition, except the contents and source of information, remain. In addition, a sample of the report or screen and a description of it augment the original definition.

The team originates the function descriptions during the prototype definition phase and makes draft additions during prototype exercising. Those team members assigned to complete the documentation should then put the function descriptions into their final form.

The description of a system's inputs describes them, tells how to prepare and enter them, and explains what to do if problems occur in processing them. If the system uses screens for entering data, help screens may provide some of this documentation. Those who develop this input documentation should use what is available as sources. The sources include the data dictionary containing a description of the input records and the elements in them. The team members should also have drafted additional information about the inputs during exercising.

The original description of the controls is a part of the prototype definition. The team may have made draft modifications to their description during exercising.

The descriptions of the system's edits and validations should have originated during the prototype exercising. The data dictionary should contain a current description of them. If the data dictionary is accessible to the average user and easy for him to understand, nothing should need to be done explicitly for the user guide. If not, team members may need to explain the edits and validations for the users.

The help screens may repeat some material that appears in other parts of of the user guide. Help screens organized by the function the user wants to perform may be very useful. For example, a payroll system may contain help screens that explain entering information on a new employee, changing deductions, and recording manually prepared special checks. The team may also prepare help screens which a user may call up when he has trouble with a screen or an output.

TABLE 7-6 Report and Output Screen Catalog Contents

COMPONENT	SOURCE
• Purpose	• Logical definition
• Sequence	• Logical definition
• Totals	• Logical definition
• Frequency	• Logical definition
• Volume	• Logical definition
• Sample	• New
• Description	• New
• Distribution (report only)	• Logical definition
• Special forms (report only)	• Logical definition
• Relationship to other screens (screen only)	• Logical definition
• Access (screen only)	• Logical definition

Data Dictionary

The data dictionary entries for the system should be complete, having been initiated during the prototype definition and updated during the prototype exercising.

Operating Documentation

Information Systems team members should prepare operating documentation to describe how the system has been installed and the instructions for running it. Those who will operate the system, presumably members of the computer operations organization, will use the documentation. The contents of the operating documentation are peculiar to the operating system the system will run under and the hardware it will run on. Organizational standards should specify the contents and organization of this documentation. A programmer-analyst from the project team should lead its preparation.

Program Documentation

The program documentation describes the programs that comprise the system. A better title for it might be "Modification Documentation" since it is what is needed by programmers to maintain the system. An Information Systems programmer-analyst from the team should lead the program documentation effort.

Developing program documentation is ensuring that all the program viewpoints can be followed. The viewpoints that program documentation gives are its most important feature. The viewpoint when looking at a program is the description of the functions it performs and the information it uses and produces. The viewpoint when looking at a function description or a data dictionary processing rule is knowing what programs accomplish it. The viewpoint when looking at an information element or a record is locating where it is used or produced.

Good documentation of programs may be no more than making certain that:

- If you know the program, you can find the functions it performs and the data it works with
- If you know the function, you can find the program that does it and the data it works with
- If you know the data, you can find where it is used and where it is produced.

The data dictionary system may provide many of the capabilities needed to locate where data is used or produced and where functions are performed.

The user guide describes functions. Also, the original function descriptions in the logical definition describe processes, the outputs they produce or the part of the database they update, and the inputs they use. The programs or separate documentation of them, if that is necessary, should include references to this information and to any notes that were made as the prototype was built and exercised. These references may be no more than the use of the titles of the function descriptions located outside the programs.

The references within the programs to information elements and records, whether on inputs, outputs, or the database, should use terms that the data dictionary system can recognize and unequivocally relate to the proper description.

The data dictionary system should also compile a where-used list that will enable it to find all uses of any information element or record.

The system flow diagram, which was done as part of the physical prototype definition, is another succinct set of pointers to the various program modules. Many Information Systems persons rely on the system flow for a quick introduction to the physical structure of a system.

The team members documenting the programs should make sure to update it.

The program documentation should include documentation of the test system and how to use it. It may also include such things as program structure charts.

OPERATING ACCEPTANCE

Operating acceptance is the verification by the computer operations organization that a system and its associated documentation conform to the standards for an operational system. A programmer-analyst team member should lead the operating acceptance effort.

The work to be done is peculiar to each organization and its standards. A request is usually made to the operations organization to accept a system. The operating documentation of the system with various statements and demonstrations that requirements have been met, for such things as storage allocations for the database, accompany the documentation.

A representative of the operations organization reviews what the project team submits. Operations will accept it when it conforms to standards.

CONVERSION

Conversion is transferring data from an existing system to a system that replaces it. It may be manual, partly manual and partly computer system, or all computer system. Users will need to review the results of the processing, especially the edits, validations, and controls, and correct any problems that are detected. A user team member should be primarily responsible for doing it. If much manual effort is required, users from outside the project team may help to do it. If at least part of it is done with a specially developed computer system which has not been installed as part of the organization's production systems, Information Systems persons may need to help run the programs.

You should convert data from the existing system selectively. You may not need to convert all of it before beginning to operate the new system.

The definition of the conversion process should specify that the data from the existing system will be processed through the edits, validations, and controls of the new system. This ensures controlling the conversion and discovering any possible incompatibilities between the two systems. Since the conversion process is exercised as is the rest of the prototype, any such problems should have been corrected before implementation.

If the new system has on-line input, the conversion process may simulate this with batch input from the existing system.

The team members should begin conversion as soon as possible after the beginning of the implementation phase and complete it before the completion of the system installation.

INSTALLATION

Installing the system is turning it on. When the team has completed the volume and parallel testing, performed the formal system acceptance, finished the system documentation, secured operations acceptance, and converted the data, they may install the system. Installation primarily involves coordinating the beginning of use of the system by the users, Information Systems, and computer operations.

With installation, the prototype becomes the operational system.

SUMMARY

1. Implementation is the final documentation, testing, acceptance, training, converting, and installing needed to make the system operational.
2. Overlap implementation tasks to complete the phase quickly.
3. Continue team reviews and maintain version control.
4. Volume testing checks those facets of the system that are sensitive to large volume.
5. Parallel testing compares the processing of the same data through both the new system and the existing system.
6. System acceptance is the verification that the system meets the user requirements and the design criteria.
7. System documentation describes a system and its history.
8. Operating acceptance is the verification that a system and its associated documentation conform to the standards for an operational system.
9. Conversion is transferring data from an existing system to a system that replaces it.
10. Installing the system is turning it on.

Chapter Eight

ENHANCING AND MAINTAINING THE SYSTEM

This chapter contains a few comments on modifying a system that was originally developed with the prototyping methodology. In contrast to this chapter, the description in the earlier chapters of developing a prototype to replace all or part of a system presumes the system being replaced may not have been developed with prototyping.

If you follow the methodology described in the previous chapters, you may design and develop the prototype, and install the system it became. The new system will not be static, though. Someone will start a change request for it before the users have rumpled the pages of the first reports it produced. The requested change will be to fix some part of the system that does not work correctly, or change its processing, or replace the system, or eliminate it. Table 8-1 lists the types of system changes.

INITIATING A CHANGE

The first things that need to be done with a request for a system change are classifying it properly and securing approval for any project it may require. Determining whether a change is maintenance or an enhancement is often especially difficult.

Definitions of enhancements and maintenance should clarify this brief discussion.

TABLE 8-1 System Changes

- Enhancement
- Maintenance
- Replacement
- Elimination

Enhancement of a system is changing part of it, adding features to it, or eliminating features from it.

Maintenance of a system is correcting processing that does not conform to the system requirements.

Users and Information Systems persons often spend considerable time and effort arguing whether a particular request is an enhancement or maintenance, especially if costs are allocated differently depending on which a change is. If a system is failing, fixing it is clearly maintenance. But many things users would like to call maintenance may actually be enhancements. If the system does not do what the requirements say it should do, making it conform is maintenance. If what the user wants the system to do is not described in the requirements, modifying it is an enhancement.

A request for a system change may begin as a service request, as described in Chapter 2, or it may go to a group responsible for responding to system problems.

PROBLEM RESOLUTION

Resolving a system problem includes the steps listed in Table 8-2. Logging it in a problem log, and possibly assigning a control number to it, helps ensure that it is resolved and not lost.

An analyst should investigate the problem next. It may only require an explanation; it may be no problem at all; it may require an enhancement; it may require a change to the documentation; or, it may require a change to the system itself.

The analyst may conclude the investigation by defining a solution for the problem. Implementing the solution may require a project,

TABLE 8-2 Steps in Resolving a System Problem

- Logging
- Investigating
- Defining a solution
- Establishing a project
- Planning and scheduling
- Implementing solution
- Resolving the problem

whose work will have to be planned and scheduled. Performing the project should implement the solution.

An important final step, whether a project was required or not, is to secure the user's agreement that the problem has been resolved. An entry showing its resolution should close the problem log for each problem.

MAKING A CHANGE

Changing a system developed with prototyping should be easier than changing one developed other ways. Prototyping produces materials that give you an advantage. From it you have the system documentation, the test system, version control of the system components, and the data dictionary entries.

Even if the change requested is to eliminate the system, the materials developed during prototyping will at least provide a good history of the system.

If the change is to replace the system, you should develop the replacement system using the prototyping methodology, as the earlier chapters describe it.

In the Prototyping Methodology, modifying a system, either by enhancing or maintaining it, is continuing the prototyping process.

RESPONSIBILITY FOR SYSTEM CHANGES

In some organizations a remnant of the original project team may remain to do system changes. In other organizations the responsibility, especially for maintenance, passes to a separate group that specializes in maintenance. In still other organizations ad hoc groups from either a development or maintenance group do system changes, whether they are replacements, enhancements, or maintenance.

USING THE DATA DICTIONARY

The data dictionary is especially important for modifying a system, because *a system change is a change to one or more information elements*—how they are processed, their size, characteristics, validations, sources for posting, records they are kept in. You either modify existing elements or add new ones.

After you boil a change down to the elements it affects, you need to find and to change all the occurrences of the elements. The where-used feature of your data dictionary system should provide this capability. What you need to know is:

- Affected elements
- Alias names for them
- All the uses of the elements and their aliases.

If a change only involves adding new elements to a system, you probably do not need the where-used feature of the data dictionary system. But you will still want to use the data dictionary to add the new elements.

USING TEST SYSTEM

Having a test system available makes checking changes much easier than when no test system is available. When the test system already contains data to test the feature that is being modified and no changes to the data are required, the test data sets may be run against the modified programs. Checking the results then should be easy.

If you have to add or modify the contents of one or more test data sets to test the change, being able to start with existing test data sets still will make the work easy.

After you have tested a change, you may have modifications that you want added to the test data sets. You should give them to the person responsible for maintaining the test system.

CHANGE CONTROL

If you did not already have version control of system components beforehand, you should have installed it to support your prototyping project. Using it will help you to modify the system.

SUMMARY

1. The types of system changes are:
   - Enchancement
   - Maintenance

- Replacement
- Elimination

2. Prototyping materials that help you make changes include:
   - System documentation
   - Test system
   - Version control of the system components
   - Data dictionary entries
3. An enhancement changes the requirements; maintenance makes the system conform to the requirements.

BIBLIOGRAPHY

Allen, Joseph & Lientz, Bennet P., *Systems in Action.* Santa Monica, CA: Goodyear Publishing Company, Inc., 1978.

Andrews, William C., "Prototyping Information Systems." *Journal of Systems Management,* 1983, *34*(9), 16–18.

Asner, Michael & King, Alan, "Prototyping: A Low-Risk Approach to Developing Complex Systems," *Business Qtrly (Canada),* 1981, *46*(3), 30–34.

Batt, Robert, "Systems Development Change Seen Inevitable," *Computerworld,* 1983, *17*(22), 31.

Boar, Bernard H., *Application Prototyping: A Requirements Definition Strategy for the 80's.* New York: Wiley Interscience, 1984.

Booth, G. M., *The Design of Complex Information Systems,* New York: Mc-Graw Hill, Inc., 1983.

Burkhalter, Dick, "Warning to Anyone Moving COBOL Applications to Micros," *Computerworld,* 1984, *18*(51), ID/1.

Chapin, Ned, "Prototyping—Quick, Not Dirty . . ." *Data Management,* 1983, *21*(10), 46–48.

Fedanzo, Anthony Jr., "Running Your COBOL on Micros." *Computerworld,* 1984, *18*(47), ID/1.

Frank, Werner L., "Structured vs. Prototyping Methodology," *Computerworld,* 1983, *17*(33), 51–52.

Gane, Chris & Sarson, Trish, *Structured Systems Analysis: Tools and Techniques.* Englewood Cliffs NJ: Prentice-Hall, Inc., 1979.

Gilhooley, Ian, "System Development Mythology," Datamation 1983, 29(6), 272–276.

Gremillion, Lee L. & Pyburn, Philip, "Breaking the Systems Development Bottleneck." Harvard Business Review, 1983, 61(2), 133.

Johnson, L., "User Benefit Studies." Telecom Australia, New Zealand Interface July 1983, pp. 64–66.

Kruesi, E., "The Human Engineering Task Area." *General Electric Company Computer*, 1983, *16*(11), 86–93.

Lantz, Kenneth E., "Surviving the Computer Explosion of the Eighties." *The Insurance Accounting and Systems Association Interpreter*, 1983, *42*(10).

Lord, R., Miller, J., & Kahane, R., "A Procedure for the Estimation of Software Development Costs." *IFAC/IFIP Workshop on Real time Programming 213–27*, Oxford, England: Pergamon, 1978.

Mason, R. E. A. & Carey, T. T. "Prototype Interactive Information Systems." *Communications of the ACM*, 1983, *26*(5), 347–354.

Marett, W., "New Development Tool Project Based On 'C'." *New Zealand Interface*, July 1983, 16–17.

McCallam, D. H., "A Comparison of Three Software Costing Techniques." *Proceedings of the IEEE 1983 National Aerospace and Electronics Conference*, 1983, *2*, 1021–6.

McCracken, Daniel D., "Software in the 80s: Perils & Promises." *Computerworld*, 1980, *14*(38), 5–10.

Meyer, K. & Kovaco, A., "Model Systems." *Datamation*, 1983, *29*(9), 248, 250, 252.

Nocentini, Stefano, "The Planning Ritual." *Datamation, 1985, 31*(8), 122–128.

Pope, K. S., "Systems Analysts and Cultural Lag." *Australia Computing Journal*, 1979, *11*(1).

Rubin, H. A., "Macro-Estimation of Software Development Parameters: the Estimacs System." *Softfair: A Conference on Software Development Tools, Techniques, and Alternatives, Proceedings 109–18.* Silver Spring MD: IEEE Comput. Soc. Press, 1983.

Sarvari, I. L., "The Case for Prototyping." *Canadian Datasystems*, 1983, *15*(10), 100–104.

Shelly, Gary B. & Cashman, Thomas J., *Business Systems Analysis and Design*, Brea CA: Anaheim Publishing Company, 1975.

Shoor, Rita, "Panel Backs Human-Engineered Software Tools," *Computerworld*, 1980, *14*(22), 36.

Siegers, W., "Prototyping." *Informatie* (Netherlands), 1984, *26*(1), 64–69.

Watson, Hugh J. & Carroll, Archie B., *Computers for Business: a Managerial Emphasis.* (Rev. Ed.). Dallas TX: Business Publications, Inc., 1980.

Websters New Collegiate Dictionary. Springfield MA: G. & C. Merriam Co., 1977.

Wetherbe, James C., "Systems Development: Heuristic or Prototyping?" *Computerworld,* 1982, *16*(17), SR14-15.

Wirth, Robert, "Future Direction of Software Technology." *GTE Automatic Electric Worldwide Communications Jrnl,* 1983, *21*(3), 70–75.

Whitlock, Brad, "Prototyping May Alleviate DP Crunch." *Computerworld,* 1982, *16*(24), 71, 75.

Zajonc, Peter & McGowan, Kevin, "Proto-Cycling, New Languages Offer Cost-Effective Programming." *ComputerData* (Canada), 1983, *8*(1), 5.

INDEX

D

E

T